CHRIS PACKHAM'S
Nature Handbook

Explore the wonders of the natural world

Penguin Random House

REVISED EDITION

Senior Art Editors Ina Stradins, Pooja Pipil
Senior Editors Rob Houston, Dharini Ganesh
Editor Shambhavi Thatte
Editorial Consultant Tim Harris
Picture Researcher Laura Barwick
Managing Editors Angeles Gavira Guerrero, Rohan Sinha
Managing Art Editors Michael Duffy, Sudakshina Basu
DTP Designers Nand Kishor Acharya, Anita Yadav
Senior Jacket Designer Suhita Dharamjit
Jacket Designers Stephanie Tan, Tanya Mehrotra
Production Editor Kavita Varma
Production Controller Laura Andrews
Pre-production Manager Balwant Singh
Production Manager Pankaj Sharma
Jacket Design Development Manager Sophia MTT
Editorial Head Glenda Fernandes
Design Head Malavika Talukder
Associate Publishing Director Liz Wheeler
Art Director Karen Self
Pubishing Director Jonathan Metcalf
Design Director Phil Ormerod

FIRST EDITION

Senior Art Editor Maxine Pedliham
Project Editor Ruth O'Rourke-Jones
Editors Martha Evatt, Laura Palosuo, Cressida Tuson,
Miezan van Zyl, Jamie Ambrose, Lara Maiklem
Designers Rebecca Tennant, Silke Spingies, Simon Murrell,
Stephen Knowlden, Elaine Hewson
Illustrator Dan Cole/The Art Agency
Production Editor Tony Phipps
Production Mandy Inness
Creative Technical Support Adam Brackenbury, John Goldsmid
Jacket Designer Duncan Turner
Picture Researcher Laura Barwick
Managing Editor Sarah Larter
Managing Art Editor Michelle Baxter
Reference Publisher Jonathan Metcalf
Art Director Phil Ormerod

This edition published in 2022
First published in Great Britain in 2010 by Dorling Kindersley Limited
DK, One Embassy Gardens, 8 Viaduct Gardens, London, SW11 7BW

The authorised representative in the EEA is Dorling Kindersley Verlag GmbH.
Arnulfstr. 124, 80636 Munich, Germany

Copyright © 2010, 2022 Dorling Kindersley Limited
A Penguin Random House Company
2 4 6 8 10 9 7 5 3 1
001-327085-Mar/2022

Foreword and chapter introductions copyright © Chris Packham.
A CIP catalogue record for this book is
available from the British Library.
ISBN: 978-0-2415-3928-6
Printed in the UK

For the curious
www.dk.com

MIX
Paper from responsible sources
FSC™ C018179

This book was made with Forest
Stewardship Council™ certified paper -
one small step in DK's commitment to a
sustainable future. For more information
go to www.dk.com/our-green-pledge

DISCLAIMER
Always remember to keep safe and be sensible when exploring an unknown terrain.
The Publisher has set out some basic guidelines on safety on pages 40–41, but it is
the responsibility of every user of this book to assess the individual circumstances
and potential dangers of any habitat they wish to explore. The Publisher cannot
accept any liability for injury, loss, or damage to any user following suggestions in
this book.

The Publisher would draw the reader's attention to the following particular points:
• plants may be poisonous or protected by law from picking or uprooting
• fungi and berries should only be collected for consumption at reader's own
risk since many fungi and some berries are poisonous
• wild animals may bite and/or sting – take suitable precautions and a first aid kit.

Consultant Editor

Chris Packham developed a fascination with wildlife from an early age and studied zoology at Southampton University, England. He has written several books on wildlife and has hosted many nature-based TV shows for the BBC, including *Springwatch* and *Autumnwatch*. Chris is involved with many wildlife conservation organizations including The Wildlife Trusts, The Wildfowl and Wetlands Trust, The Bat Conservation Trust, and is a Vice-President of the RSPB.

Contributors

Steve Backshall (Mountain and hillside) is a naturalist, author, and television presenter. He has travelled to more than a hundred countries, discovered new species, and climbed some of the world's highest mountains.

David Chandler (The web of life; Lake, river, and stream) is a freelance writer and environmental educator. David's books include the *RSPB Children's Guide to Bird Watching*, *All About Bugs*, and *100 Birds to See Before You Die*.

Chris Gibson (Coast) is a naturalist who writes, teaches, and broadcasts about the natural world. He is a senior specialist for Natural England and the author of *RSPB Pocket Nature Seashore*.

Robert Henson (Weather) is a meteorologist and science journalist based in Colorado, USA. He has worked as a tornado researcher and written extensively on climate change.

Rob Hume (What a naturalist needs; Forest; Consultant) worked for the RSPB for 30 years, editing *Birds* magazine for 15, and has written around 30 books, including DK's *RSPB Complete Birds of Britain and Europe* and *RSPB Birdfeeder Guide*.

James Parry (Scrubland and heath; Grassland; Desert) is a writer and lecturer who has travelled widely to study wildlife and different habitats. He has written several books on natural history.

Dr Katie Parsons (Close to home; Farm and field) has a PhD in animal behaviour and ecology. She currently lives on a 30-acre farm in the Devon countryside, which she manages for both wildlife and holidaymakers.

Elizabeth White (Tundra and ice) is a documentary film-maker for the BBC Natural History Unit. She has a PhD in animal behaviour and has filmed wildlife across the globe, including in the Antarctic and the high Arctic.

Contents

Foreword
by Chris Packham 6

The web of life 8

Weather and sky 18

What a naturalist
needs 34

Close to home 46

Farm and field 66

Forest 80
↣ Deciduous woodlands 82
↣ Coniferous forests 106

Scrubland
and heath 118

Grassland 130

Mountain
and hillside 140

Lake, river,
and stream 156

Coast 174
🐚 Beaches 176
🐚 Cliffs 194
🐚 Coastal wetlands 202
🐚 Ocean 208

Tundra and ice 210

Glossary 218

Index 220

Acknowledgments 224

ABOUT THIS BOOK

This book is intended to be an inspirational guide to exploring, understanding, and observing the natural world, wherever you may be in Britain or Europe. The species included are examples of the type of wildlife that exists in each habitat, wherever they occur in Britain or Europe. Not all examples shown for a given habitat will be found together or in one specific geographical location.

Foreword

I wonder how many species are living on our world… right now? What would it be like to see through the eyes of a butterfly. How does it feel to be a swallow, swooping over a lake? How big is – or was – the very largest giant squid? What did a tyrannosaur really sound like?

Let's be honest, I'll never know the precise answers to these questions, but there isn't one of them that I couldn't come up with an educated theory about. With a little research I could "guestimate" some of them. This book is about exciting and satisfying those with a fascination for the natural world, those who, like me, want to try to understand what makes it all tick, how that ticking takes place, and whether there's a proper name for it. When I was the eager 10 year old trying to understand everything about animals and their habitats, pawing the pages of my animal encyclopaedias, ricocheting from snakes shedding their skins to the "miracle" of metamorphosis, what I really wanted was detail. I needed "sloughing" skins rather than "shedding" and I already knew that while the processes playing out in the chrysalis were hidden from me – and an explanation of them irritatingly absent from my book – there was certainly no miracle involved. What annoyed me most of all was that I was sure that there was a white-coated scientist out there who knew all the answers. Perhaps it's because of this legacy that I've always refused to "dumb down" information. I never did "mini-beasts", but stuck with invertebrates instead. If children so easily and ably enjoy the names of dinosaurs then why shy from teaching them a proper naturalist's vocabulary? Young doesn't mean simple – it means hungry and absorbent – and if people develop a fascination for the natural world at a later age, then in these times of wider awareness and access to all information they are well prepared to understand properly explained science and the workings of the natural world.

And that's what I wanted this book to be, a beautifully presented, full-colour, 21st-century look at the workings of the natural world for all the family. I wanted to learn a lot from it myself and to update things I'd read long ago, and

I also wanted it to facilitate a wider ambition – one that for all my swotty obsession with everything that slithers, sneaks, slimes, and stings, wasn't realized until I was in my twenties. I wanted it to portray the bigger picture, to offer an opportunity to realize how and why all of the little things are linked, to illustrate aspects of an invisible web that necessarily joins all our planet's life. And through this to present the basis of a framework to support the readers own knowledge or observations, so that these could fuse recognitions of relationships between the infinite multitude of seemingly separate species, their physiologies, ecologies, and behaviours.

But of course you won't achieve this solely by reading about it or looking at the pictures – you have to get out there where the action is, and this book aims to inspire you to do just that. That action doesn't have to be "big" or "glamorous" – and it certainly doesn't have to be exotic. There are incredible dramas playing out beneath logs in your garden, life and death battles being fought in your local park, and epics unfolding on wastelands or nature reserves around your corner. And discovering them for yourself is always better, always more rewarding.

I hope that with any such deeper understanding comes a deepened desire to protect and conserve those truly perfect things in our world and again a realization that all such efforts will be wasted unless we immediately employ our armoury of technologies to combat and adapt to our planet's changing climate. From twinkling stars and new clouds, to tadpoles and their metamorphosis, there is so much pure wonder portrayed on these pages – do we really want to destroy it? No. So let's empower ourselves to make a difference where and whenever possible.

This book is dedicated to my Mum, who allowed my foxes to poo on her carpets without too much complaint. I would also like to extend a gargantuan thank you to all the authors who have so diligently researched and skilfully written their sections to eloquently explain everything from the minutiae to the massive; to the team of editors at DK who have

carefully melded these gems together to produce an effective synergy; and to the designers, photographers, and artists who have given such stunning visual life to all the ideas and explanations. I so wish I was 10 and starting to explore the natural world again!

Foreword to the 2022 edition

Since writing this more than 10 years ago I have changed. Because the world has changed. I'm now 60 years old and without being morbid I have to recognize that the time I have to do my small bit to make the world a better place, environmentally and ecologically, is shorter. I have to work much harder. And sadly it's much harder to do so, because the problems that should have been addressed back then, and before, have not been adequately sorted and many things have gotten worse – less habitat, fewer organisms, less time to protect them. We've been tinkering and making some progress, but we've been distracted too, so now we must turn our attention away from the critical but little issues and confront the critical but massive issues of climate breakdown and biodiversity loss.

Sounds impossible, doesn't it? How can I, how can you seriously do anything meaningful when it comes to addressing these existential challenges? Well, it's simple. We start small and we grow big. Step one is singularly the most important – each and every one of us has to empower ourselves to make a difference. As an individual, just me and you. There is obviously so much there to do and I don't need to make a list here, just look at yourself, your life, and lifestyle and ask yourself, "What could I change to make things better?" Then do it. Then feel good about it and, most importantly, tell others what you have done and see if they too can change. Be tolerant and patient though – not everyone will begin this journey at the same time, progress at the same pace, or choose to do the same things. Be kind, it's better to applaud a small positive change than to deride a large amount of inaction.

When people do change with us, we start to build communities, and the difference being made gets magnified. It starts to become meaningful.

Ultimately, however, we need our decision makers, those we have elected to represent us, to act on larger scales. Unfortunately, but predictably, they have just wasted an enormous opportunity to really start getting the job done: COP26 (26th UN Climate Change Conference of the Parties). They are not going to fix it, so it's down to us to drive those changes and that's what we must do – in a polite, peaceful, and democratic way.

We are not short of solutions, we have evolved the technologies and capacities to repair, restore, regenerate, reinstate, reintroduce… and we have tested and perfected this portfolio of tools. We just need to roll them out more broadly and more rapidly. A lot more urgently…

Some of the people this book will hopefully inspire will inherit a very different world than the one I grew up in, with those ladybirds, tadpoles, caterpillars, fox cubs, and grass snakes always just within my reach, all that fascination just waiting for my curiosity to nudge. I hope that they still become inspired, still light up when they encounter nature, that a spark ignites, which burns to fuel their own desire to nurture nature. Because without that energy we will be in trouble. Please give or show or read this book to a child. Please help sow the seeds of a better future for all life on our wonderful Earth.

Chris Packham

The web of life

The simple beauty of life can be relished on many levels. A single
bright-red ladybird on a fingertip is perfect. The fresh scent of a
rose is sublime. The tiny rainbows seen flashing from the wings
of aphids on the rose's stem are also unexpected gems, and the
marvel of a myriad of ants flying up into the summer sky makes
an urban spectacle. Each is individually remarkable, but then so
are the relationships that essentially and intrinsically link them
all. There is an undeniable and satisfying beauty to be found in
an understanding of these webs that knit life together.

The nature of the planet

Much of the time, we are aware only of life immediately around us, yet this is only a small part of a much larger network. Life on Earth exists in many places – some very different to others, but all are connected.

The thin green line

Life in all its forms is found almost exclusively on the Earth's outermost layers, including the land, oceans, and the atmosphere surrounding the planet. This narrow strip is known as the biosphere – a word that literally means "life ball". Within it are millions of species, of which humans are one, with each dependent on others for their survival. The biosphere isn't uniform, however – it is a collection of different, yet interconnecting habitats, which have many ill-defined boundaries between them.

TUNDRA
Exposed, cold, and treeless, with many lichens and mosses, tundra is a habitat of the far north.

Key

- Grassland
- Desert
- Tropical forest
- Temperate forest
- Coniferous forest
- Mountains
- Polar regions and tundra
- Rivers and wetlands
- Coral reef
- Oceans

WORLD BIOMES DISTRIBUTION
The scientific word for a habitat is a biome. This map shows the variety of these biomes and their distribution, which is determined by climate and geology. Human impact on the environment isn't indicated – areas shown as temperate forest, for example, may now be farmland.

GRASSLAND
Grassland includes savannas, steppes, and prairies. It experiences more rainfall than deserts, but is drier than forests.

REEDBEDS IN NORFOLK, UK
Many of these important habitats would be lost today if they were not periodically managed.

HABITAT-MAKER
Left to their own devices, some habitats are transient, changing from time to time. Reedbeds are a good example. Often, dead vegetation builds up at the base of the reeds. This dries out the reedbed, allowing other species to gain a foothold. Scrub may take over, and ultimately woodland, which is a much more stable habitat.

AQUATIC
Aquatic habitats include lakes and streams to rivers and oceans. They may be saltwater or freshwater.

More than one home

Some animals have a very strong connection with a single habitat – Europe's bearded tits, for example, are small birds found mainly in reedbeds. Other species make themselves at home in many habitats – the adaptable carrion crow can be seen in woods, uplands, and foraging on estuaries, among other places. Dragonflies make a big habitat change when they become adults.

adult winged dragonfly emerging from its larval "skin"

TRANSFORMER
The first part of a dragonfly's life is spent underwater as a larva, yet once it matures, it becomes an aerial predator.

LIFE ON EARTH

All life on Earth exists as part of an intricate web of interconnections. These images help to put some of these into context. They start with an individual of one species, and, step by step, move on to the biosphere. Individuals of any species don't generally live in isolation – others of their kind normally reside in the same area. Together, these make up a population. Add populations of other species in the same area and this builds into a community. The community lives in a specific habitat, with a certain climate, geology, and soil – together these living and non-living components make up an ecosystem. Put all the ecosystems together and you have the biosphere. In this way life on Earth is interconnected, and we should take care to not tip the balance.

FOREST
Forests are highly varied and species-rich habitats. Types of forest include northern boreal, tropical, and temperate forests.

Find your own biome on the map. Perhaps it was once temperate forest.

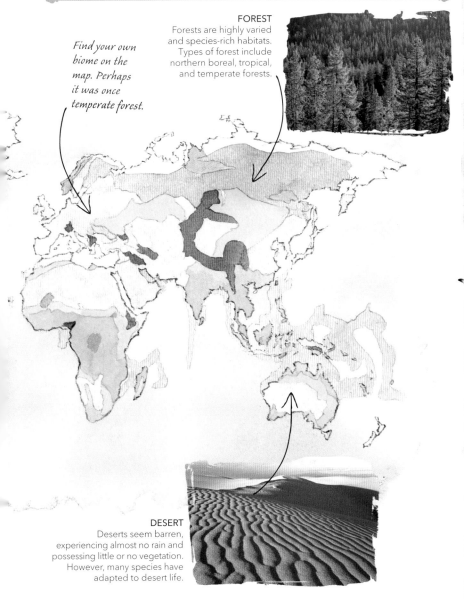

DESERT
Deserts seem barren, experiencing almost no rain and possessing little or no vegetation. However, many species have adapted to desert life.

INDIVIDUAL
As a naturalist, you might encounter just one individual of a species. However, it is part of a larger group.

POPULATION
The individuals of a species in one area make up the population. Different species have different sized populations.

COMMUNITY
All the populations together form a community, in which population fluctuations in one species have an impact on other species.

ECOSYSTEM
Ecosystems may be large or small, and combine living components with an area's physical characteristics.

BIOSPHERE
This is the "ball of life". It is made up of all individuals in every population in every community and all habitats on the planet. The true worldwide web.

The diversity of life

The diversity of life on Earth is extraordinary. As a naturalist, there is always something new to understand, experience, and enjoy.

Scientists have identified about 1.8 million species, and it is estimated that as many as 6 to 12 million more are waiting to be discovered. Humans are just one animal species among many, but we have a unique role to play in understanding and conserving the rest.

Evolution

Just as human families exhibit variations in, for example, eye colour, animals vary within a species. As differences are passed on to subsequent generations, species may slowly evolve into individuals with varied appearances and capabilities. For example, if one bird has a larger beak than its neighbour and is better at feeding its young, more of them survive. Some of its chicks will inherit larger beaks, and, with time, more offspring acquire larger beaks until they look quite different to their smaller-beaked relatives. If there comes a time when the birds with large beaks can no longer breed successfully with the birds with small beaks, a second species has been created.

SLOW PROGRESSION
The elephants we recognize today are believed to have evolved from a prehistoric animal called *moeritherium* – an animal that more closely resembled modern tapirs.

MAMMALS
Mammals make up around 6,400 species, including this red squirrel, tiny bats, massive whales, camels, kangaroos, polar bears, cheetahs, giraffes – and humans.

BIRDS
More than 10,000 known bird species are hugely diverse, ranging from ostriches to penguins, albatrosses to eagles, ducks to owls, hummingbirds, and sparrows.

REPTILES
These are cold-blooded vertebrates and their bodies are usually covered in scales. There are probably more than 10,000 known species, including lizards, snakes, turtles, and crocodiles.

AMPHIBIANS
These animals have adapted to life both in water and on land. There are more than 8,000 species of amphibians, including caecilians, salamanders and newts, and frogs and toads.

FISH
Earth's fresh, brackish, and salt waters are home to almost 31,000 known fish species, including salmon.

INSECTS
Insects are the most abundant class of animals on Earth. More than 1 million insect species share the planet with humans – over 350,000 of them are beetles.

FLOWERING PLANTS
Around 260,000 flowering plant species have been recorded, on land and in water. These include grasses, trees, and wildflowers, such as these harebells.

TREES
The definition of what is considered a tree is not absolute, but there are an estimated 100,000 tree species in the world.

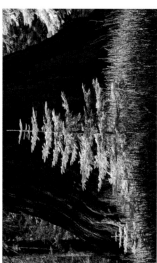

FUNGI
More than 120,000 species of fungi have been described. Toadstools and mushrooms belong to this group.

EVOLUTION IN ACTION
The five digits in this skeletal paw, and what looks like a thumb, belong to the giant panda, a member of the bear family. The "thumb" is actually a wrist bone, but it is much larger than that of, say, a brown bear. It can also move, is padded, and works with the true digits to make it easier for the panda to handle bamboo, its preferred food. This appendage may have evolved over thousands of years as a trait that was beneficial to the panda's survival.

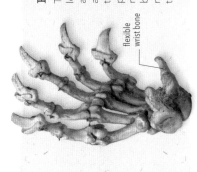

flexible wrist bone

Adaptations
Evolution is about change, and if an inherited characteristic increases the chance of survival by making an animal better at finding food or avoiding predation, for example, then those attributes are more likely to be passed on to the next generation. Within the animal kingdom, some species have – over many generations – evolved an array of adaptations to meet the challenges of life, including camouflage, super-sharp senses, mouthparts that function as specialist feeding tools, anti-freeze in the blood, or even feathers that hold water.

CORAL SNAKE

red touches black bands, not yellow

SCARLET KINGSNAKE

MIMICRY
Predators may keep their distance from some non-venomous species of kingsnake, which have evolved to resemble highly venomous coral snakes.

PERFECTLY ADAPTED
Sword-billed hummingbirds use their ultra-long beaks to reach nectar in flowers, pollinating them in the process.

13

Animal life

Animals occupy particular niches within the complex web of life, and have evolved various strategies and behaviours to ensure survival.

SCAVENGER

Herbivores, carnivores, and scavengers

Simply put, green plants use the Sun's energy to grow, herbivores eat the plants, and carnivores eat herbivores. But feeding relationships are often more complex. Carnivorous foxes prey on herbivorous rabbits, but also eat fruit. Crows may scavenge from the dead bodies of both animals, but also eat seeds, fruit, insects, and small animals. They are taking advantage of evolutionary niches by developing different eating habits.

HERBIVORE

CARNIVORE

Early bird or night owl?

Not all animals are active at the same time, which can reduce competition between species; for example butterflies take nectar from flowers during the day, while most moths do so at night. Animals that are active during the day, such as most lizards, are "diurnal" and those that are active at night, such as hedgehogs, are "nocturnal" (see pp.54–55). Some animals are "crepuscular", which means you are most likely to see them at dusk and dawn.

FLEXITIME

The snowy owl is a crepuscular hunter that raises its young on the Arctic tundra, where, in the very far north, there is no darkness for months during the summer. At this time of year, while the female is brooding the young, the male can have sole responsibility for feeding up to 11 youngsters, the female, and himself. To do this, he adapts his usual habits, hunting in the day.

NIGHT AND DAY
Hedgehogs are nocturnal mammals found in Europe, Africa, and Asia. Lizards can be diurnal or nocturnal – this viviparous lizard is diurnal.

DIURNAL

NOCTURNAL

Getting away from it all

Mammals need to eat to stay warm, but in winter food can be hard to find. Some survive by hibernating. During this time their metabolisms (internal processes) are turned down to a minimum, and they use the fat deposits they laid down while food was plentiful to fuel this low energy winter existence. There are also invertebrates, reptiles, and amphibians that hibernate. Migration is a strategy that is employed most visibly by some birds, but also by fish, butterflies, moths, and land and sea mammals. These animals travel huge distances, often along well-defined routes, in search of food and breeding grounds.

LAND MIGRATION
Migrating reindeer can travel more than 5,000km a year, crossing water if necessary. No other land mammal covers such a distance.

Migrating humpbacks can be seen from locations on the Pacific coast.

EPIC JOURNEY
Humpback whales migrate further than any other mammal. Their journey, between the Central American Pacific and the Antarctic, is over 8,000km.

HOW MIGRATING BIRDS NAVIGATE

A bird's ability to navigate between breeding grounds and wintering areas, which can be thousands of kilometres apart, is staggering. Visual clues assist them, for example a river may keep them on track, and the Sun acts as a compass, with birds using their "internal clock" to compensate for its apparent movement. At night they use the stars as a guide. Birds can also detect the Earth's magnetic fields and use these to navigate. As they get closer to their destination smell may help: petrels, for example, find their burrows by smell.

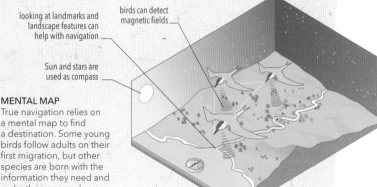

looking at landmarks and landscape features can help with navigation

birds can detect magnetic fields

Sun and stars are used as compass

MENTAL MAP
True navigation relies on a mental map to find a destination. Some young birds follow adults on their first migration, but other species are born with the information they need and make the journey alone.

Back from the brink

Human intervention in the natural world can have a dramatic impact on the lives of animals and plants.

Humans can have a detrimental effect on animal and plant populations through a variety of means. However, we also have the capacity to turn things around, and in some cases this has happened. The sea otter is one such example. Once hunted to near extinction for their fur, now, thanks to successful conservation initiatives, they can again be seen in waters off North America's Pacific coast. Conservation projects have also helped the American bison, after hunting decimated herds that once totalled many millions. The osprey, which, by 1916, had been persecuted to oblivion in the UK by egg collectors and hunters has also seen a reversal of fortune. Similarly, the large blue butterfly had disappeared from the UK by 1979, but has been successfully reintroduced. Although successes like these can be achieved, many species of plants and animals remain threatened.

OSPREY
Ospreys returned to Scotland in 1954. Round the clock protection and recent reintroduction to England has helped UK numbers rise to more than 250 pairs.

EUROPEAN BISON
The European bison has been brought back from the brink of extinction. More than 7,000 now live in protected areas like Białowieża Primeval Forest, Poland.

LARGE BLUE BUTTERFLY
Reintroduction and appropriate land management has helped save the UK's large blue population from extinction.

SEA OTTER
Reintroduction projects and legal protection have enabled populations of sea otters in the North Pacific Ocean to reach over 100,000 individuals.

Weather and sky

Perhaps no greater factor has a more important or powerful influence on all life than the weather – from the very short to the very long term. Hourly, daily, or seasonal variations exert profound effects on species and their populations, and individual events can provoke catastrophe or celebration. A cloudburst in the desert, for example, is the source of an explosion of life, but the same event could extinguish it elsewhere. The impact upon our species seems set to become ever more critical as we pitch our predictive abilities against increasingly turbulent fluctuations in the world's atmospheric conditions. Thus, understanding weather is fundamental to understanding all life on earth.

Climate and seasons

You can see variations of weather in daily and seasonal cycles and regional patterns. Together, these produce a climate: the norms and extremes that occur at a given place.

The Sun

Without solar energy there would be no climates on Earth. Daily cycles of sunshine and darkness result from the planet's rotation, and seasons are caused by the position of the Earth's hemispheres as it orbits the Sun. Because this orbit is not absolutely circular, Earth is closer to the Sun in January than in July. In about 13,000 years, however, the opposite will be true, which should warm northern summers.

POLAR CLIMATE
Some bird and whale species migrate to the polar regions as regional sea ice expands and retreats seasonally.

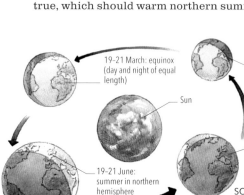

19–21 March: equinox (day and night of equal length)

Sun

19–21 December: summer in southern hemisphere

19–21 September: equinox (day and night of equal length)

19–21 June: summer in northern hemisphere

SOLSTICES
Solstices are the twice-yearly times when the Sun reaches its highest or lowest point in the sky. Due to Earth's tilt, what is summer in the north will be winter in the south.

ATLANTIC OCEAN

LIMA

Currents

Continents, sunlight, the Earth's rotation, and the Moon all influence the movement of sea water. Trade winds help drive surface water west across the tropics. Ocean currents then move warm water towards the poles in the western Pacific and Atlantic oceans, and cold water towards the equator in the eastern Pacific and Atlantic. Far more heat lies in Earth's vast, dense oceans than in its relatively thin atmosphere. This helps keep northerly London fairly mild, and equatorial Lima, Peru, surprisingly cool. Meanwhile, a broad "conveyor belt" threads through the global oceans (see map, above).

JET STREAMS
These fast-moving air currents help to regulate the climate by connecting areas of contrasting temperatures and air pressure.

OCEAN WARMTH
Despite being near Antarctica, southern Chile is insulated from extreme cold by the surrounding ocean, making it habitable for temperate-zone species.

KEY

- Polar
- Tundra
- Subarctic
- Continental
- Temperate
- Warm, oceanic
- Mediterranean
- Semi-arid
- Arid
- Subtropical
- Equatorial
- Mountain

Global zones

All habitats and biomes (see pp.10–11) are affected by climatic factors such as sunlight and moisture. The Earth is grouped into a system of climate zones (see below), with latitude (the distance from the equator) by far the strongest influence. Ocean currents and surface types (topography) are also important. Coastal deserts get little or no rain, thanks to cool offshore waters and stable air, yet thunderstorms rage across temperate zones, where heat builds more easily and air masses often clash.

IDEAL MICROCLIMATE
Huge tree canopies in tropical rainforests serve as sunscreens, keeping the air constantly warm and moist, which is ideal for animals such as butterflies and frogs.

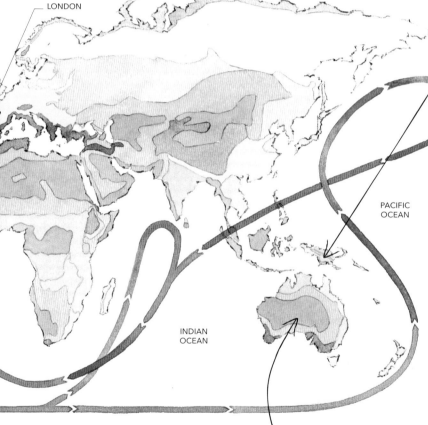

LONDON

PACIFIC OCEAN

INDIAN OCEAN

CORIOLIS EFFECT

As the planet revolves, it turns more quickly west to east in the tropics (its widest part) than in polar regions. When air currents flow from the tropics to the poles, the speed forces them to bend right over the planet's surface – a phenomenon known as the Coriolis effect. Air moving toward the equator also turns right, creating trade winds (see opposite). This effect helps explain the direction of prevailing winds and the presence of gyres.

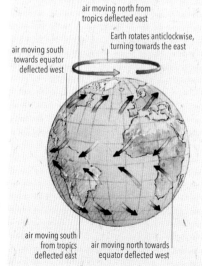

air moving north from tropics deflected east

Earth rotates anticlockwise, turning towards the east

air moving south towards equator deflected west

air moving south from tropics deflected east

air moving north towards equator deflected west

KEY

WARM SURFACE CURRENT

COLD, SALTY, DEEPWATER CURRENT

CONVEYOR BELT
Warm surface water flows from the tropical Pacific and Indian oceans around Africa, then north across the Atlantic. The water gradually sinks, forms cold bottom water, then completes the loop.

ARID CLIMATE
Despite dry conditions and large daily swings in temperature, many creatures and plants, such as lizards and spinifex grass, are well adapted to deserts.

Cloud spotting

Learning to read the sky's dazzling variety of clouds is useful for understanding air currents and can help to forecast upcoming weather.

How a cloud is formed

Water vapour is at the heart of every cloud. As warm air is forced upward, it cools, and the relative humidity increases. The rising air becomes saturated, and the water vapour collects around dust, salt, or other airborne particles to form a cloud. The type of cloud is dictated by its altitude, temperature, moisture content, and the air flow surrounding it.

CLOUD FORMATION
As water condenses in rising air, it releases heat. The heat warms the air mass, and causes it to rise further until it reaches the same temperature as the air surrounding it.

5,000m
4,000m
3,000m
2,000m
1,000m
ground level

cloud builds higher and spreads as unstable air keeps rising

condensing vapour releases heat, slowing the cooling rate

condensation adds heat and fuels rising motion and cloud growth

vapour condenses to form base of cloud as rising air expands and cools

pocket of warm air rises through cooler air

warm air rises from ground level

Identifying clouds

The higher the cloud, the lower its temperature. Some are made of ice crystals, others of water droplets, and the composition gives each a different form. Our classification of clouds is based on one created by English pharmacist Luke Howard. In 1783, intrigued by the vivid sunsets created by volcanic eruptions, he developed a cloud-naming system, presenting it to scientists in 1802. Howard divided clouds into four types: stratus (meaning "layer"), cumulus ("heap"), nimbus ("rain"), and cirrus ("curly").

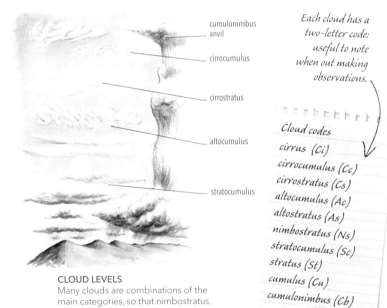

cumulonimbus anvil
cirrocumulus
cirrostratus
altocumulus
stratocumulus

Each cloud has a two-letter code: useful to note when out making observations.

Cloud codes
cirrus (Ci)
cirrocumulus (Cc)
cirrostratus (Cs)
altocumulus (Ac)
altostratus (As)
nimbostratus (Ns)
stratocumulus (Sc)
stratus (St)
cumulus (Cu)
cumulonimbus (Cb)

CLOUD LEVELS
Many clouds are combinations of the main categories, so that nimbostratus, for example, is a layer of raincloud.

SPOTTING SPECIAL CLOUDS

Some types of clouds appear rarely and only in certain areas. Noctilucent (night-shining) clouds form at heights of about 80km. Once observed only at high latitudes (in the north or south), noctilucent clouds are now being reported closer to the equator. Sometimes resembling a stack of dinner plates, lenticular clouds develop when a specific arrangement of wind layers passes over a mountain.

NOCTILUCENT CLOUD
Earth's highest cloud is most likely to be seen just after sunset or before sunrise in summer.

LENTICULAR CLOUD
With such an otherworldly appearance, lenticular clouds may be mistaken for unidentified flying objects.

High-level clouds

Forming 5–15km above sea level, high-level clouds consist mainly of sheets, patches, or streaks associated with cirrus formations. These clouds are often the first sign of an upcoming weather event, from a passing thunderstorm to a longer-lasting storm system. As sunshine or moonlight plays on cirrus clouds, ice crystals produce a range of optical effects.

1 Highly variable wind and moisture patterns can lead to a patchwork of cirrus clouds.

2 Recurring wave patterns, caused when wind blows faster above the clouds than below them, are a hallmark of Kelvin–Helmholtz cirrus clouds.

3 Contrails – narrow clouds produced by aircraft exhausts – can interact with existing cirrus clouds or spread out to form new cirrus clouds.

Medium-level clouds

The medium-level zone, ranging from around 2–5km above sea level, represents a transition region. Here, clouds take on a wide variety of shapes and sizes, affected by movements both above and below the layer as well as within it. Most clouds in this region are preceded by the prefix *alto*, a Latin term meaning "high".

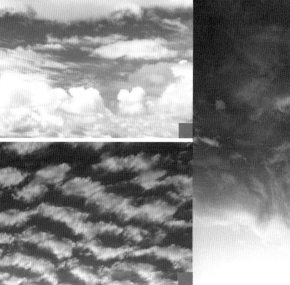

1 Cumulus clouds often push upward into the medium-level cloud zone. In this image, cumulus (bottom) clouds are joined by altocumulus (top).

2 Altocumulus often appear as vast sheets of broken cloud, especially over the ocean. Small eddies (where wind doubles back on itself) help shape these clouds into lines or arrays of cloud parcels.

3 If more moisture is present at medium levels than below, a mid-level cloud may form rain or snow that quickly evaporates as it falls, producing streaks that hang from dark clouds. These streaks are called *virga* (from the Latin for "branch" or "twig").

Low-level clouds

Warmth and moisture near Earth's surface help to make low-level clouds the most dynamic and fastest-growing. Low clouds may form when conditions are calm, which can lead to fog (see p.25). In highly unstable conditions they can set the stage for cumulonimbus clouds (thunderstorms).

1 Towering cumulus clouds extend from a smooth base upward to heights of 10km or more.

2 A vigorously developing cumulus cloud that extends to heights where the temperature is below freezing becomes a cumulonimbus, with an anvil-shaped top made up of a sheet of cirrus ice crystals.

3 Stratocumulus in the wake of a storm may appear ragged, as turbulence and wind shear – changes in wind speed or direction – eat into parts of the cloud formation.

Wet weather

After evaporating, water vapour stays airborne for a week or so. Vapour molecules condense to form clouds, before returning to Earth.

Rain

Some parts of the planet experience virtually no rain; others are deluged almost daily. How much rain falls plays a large part in the species of plants and animals that inhabit an area. Cloudy, cool areas feel more damp than their sunny, warm counterparts, which can be deceptive – on average, sunny Dallas, USA, gets nearly twice as much rain annually as cloudy London, England.

1 Frontal rain occurs as weather fronts push their way across the landscape and water condenses in the air that rises above them. Intense, frontal rain ends quickly once the front clears.

2 Orographic rain results when an air mass is forced over high terrain, such as a mountain, causing moisture to rise and condense.

3 Warm, moist air topped by cooler, drier air can lead to showers and thunderstorms that may be scattered across a summer landscape or focused along a strong front. This is called convective rain.

4 Cyclonic rain is caused by a low pressure system. Moist air spirals towards the area of lowest pressure, producing extensive clouds and precipitation.

How raindrops form

Raindrops often start as snowflakes that grow around dust motes or fungal spores within high clouds. Once large enough, they fall into warmer air and melt, turning to rain before hitting the ground. In warm climates, raindrops form without any ice being present: tiny water droplets collide, scooping up even more droplets and growing as they fall.

AIR CURRENTS
When warm air rises and cools, water vapour condenses to form clouds. When rain or hail form and start to fall, a downdraft is created by the falling precipitation.

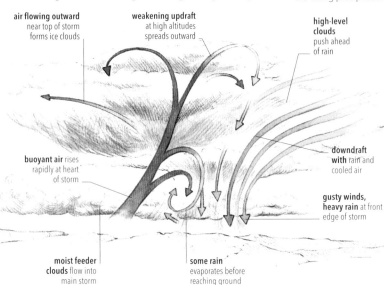

air flowing outward
near top of storm
forms ice clouds

weakening updraft
at high altitudes
spreads outward

high-level clouds
push ahead
of rain

buoyant air rises
rapidly at heart
of storm

downdraft with rain and
cooled air

gusty winds, heavy rain at front
edge of storm

moist feeder clouds flow into
main storm

some rain
evaporates before
reaching ground

Snow and hail

Many parts of the world get precipitation in frozen form. While snow develops only in clouds with temperatures that are below freezing, it may accumulate at ground level even when temperatures are slightly above freezing. Once in place, a heavy snowpack reflects sunlight and keeps the surface air cold. Hail forms when moisture-packed updrafts in a thunderstorm bring falling ice crystals back to high, cold altitudes where it accumulates, and falls as ice.

SNOWFALL
On average, 1mm of water yields about 1cm of snow. The yield is usually higher in the dry, fluffy snow of cold climates.

HAILSTONES
Sometimes as big as a grapefruit, hailstones are a spectacular and sometimes dangerous form of precipitation, causing massive damage to crops and vehicles.

SNOWFLAKES

The entrancing variety of snowflakes is due to ice crystals' tendency to grow in six-sided structures (hexagons). Different kinds usually form at different temperatures. Near-freezing conditions often lead to clusters of needles or plates. Colder air favours columns, plates, or dendrites.

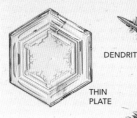

SECTOR PLATE

DENDRITE

THIN PLATE

Frost and dew

At night, especially when it is clear and calm, air near the ground can cool enough to bring the relative humidity to 100 per cent. More cooling leads to condensation on grass and other surfaces; this is either frost or dew, depending on temperature. The deposit normally disappears as temperatures rise the next morning.

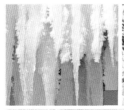

1 Rime frost, often very beautiful, is the result of water droplets that hover in below-freezing air and turn to ice when they encounter a surface.

2 Hoar frost is created when ice forms on surfaces as air close to the ground drops below freezing.

3 Dewdrops are a common sight on clear, calm summer mornings. They evaporate soon after sunrise as air warms and the relative humidity drops.

4 As surface air cools overnight, it flows into valleys and "frost hollows", where dew and frost may be especially thick.

Mist and fog

Literally a cloud on the ground, fog forms when a layer of air just above the Earth's surface cools enough so that water condenses to form cloud droplets. Even thick, dense fog may only extend a few metres above ground level. Mist is a less dense form of fog. When visibility is more than 1km the moisture is called mist, below that it is called fog.

SEA OF FOG
Cold Pacific water near San Francisco, USA, leads to frequent fogs, as moist, salt-laden air flows up the city's steep hills and engulfs the Golden Gate Bridge.

Stormy weather

Whether gentle, gusty, or gale-force, wind is the atmosphere in motion as it rushes toward and around low-pressure regions.

Breezes

You can feel the most frequently encountered winds in the form of land and sea breezes found near coasts – the result of temperature differences. Asia's monsoons are caused by a season-long pattern: summer heat warms the continent, which pulls tropical moisture inland. Localized breezes affect local habitats – and people – in very specific ways, not least by creating unique microclimates.

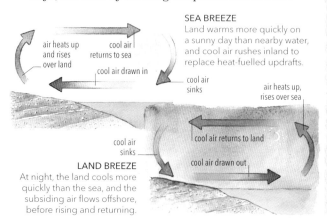

SEA BREEZE
Land warms more quickly on a sunny day than nearby water, and cool air rushes inland to replace heat-fuelled updrafts.

air heats up and rises over land

cool air returns to sea

cool air drawn in

cool air sinks

air heats up, rises over sea

cool air sinks

cool air returns to land

cool air drawn out

LAND BREEZE
At night, the land cools more quickly than the sea, and the subsiding air flows offshore, before rising and returning.

LOCAL WINDS

There are many localized winds that blow in regions around the world. The mistral, for example, is a cold northern wind that blows through southern France towards the Mediterranean Sea. As well as relentless winds, its low pressure also causes headaches in many people, and makes children and animals restless and irritable. The harmattan brings a thick haze of Saharan dust to North Africa.

Cyclones

Any area of low atmospheric pressure is, technically, considered a cyclone, although the term is usually associated with a spiralling storm. In the USA, for example, "cyclone" was once another name for a tornado, and both hurricane and typhoon are alternate names for a tropical cyclone. What we usually think of as cyclones are huge storms that generate rain, snow, and wind, and these begin as deep areas of low pressure. Winds rush in to "fill the gaps" and, due to the Coriolis effect (see p.21), begin to spiral upwards – anticlockwise in the Northern Hemisphere and clockwise in the Southern Hemisphere.

HURRICANE
Winds pull energy and moisture from warm seas. Each year 40–50 tropical storms grow strong enough to be called hurricanes, typhoons, or cyclones – all names for the same type of storm. Many cause little or no damage, but some bring extreme winds inland, causing devastation.

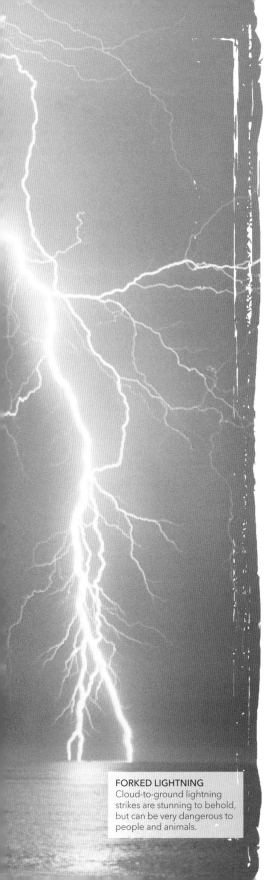

Tornadoes

The world's strongest ground-level winds occur in tornadoes – reaching speeds of 485kmph or more. These spinning columns – small, brief, but often violent – extend from thunderstorms. Tornadoes are most frequent and destructive in Bangladesh and the USA, where temperature contrasts and moisture abound, but they occur in most mid-latitude areas. Beware – if a tornado appears to be stationary but growing, it may be moving toward you.

SPIRALLING WINDS
Clouds of wind-torn debris churn around the strongest tornadoes (above). In waterspouts (left), a white ring may appear where the spinning air meets the sea.

TRACKING TWISTERS

Because "twisters" – another name for tornadoes – grow and die quickly, scientists must "chase" them to gather the data for research. Truck-mounted radar, introduced in the 1990s, has allowed scientists to profile dozens of tornadoes. Storm-chasing may look glamorous in movies and on TV, but it is mostly long, hard work. An entire season may yield only a few minutes of tornadoes.

STORM CHASERS IN TEXAS, USA

Thunder and lightning

Thunderstorms generate lightning through intense electrical fields that are produced when ice crystals and water droplets collide. Cloud-to-ground strikes are the ones that threaten people and property, but most lightning actually occurs within and between clouds. A single thunderstorm can produce many thousands of bolts in just a few hours. The intense heat generated by a lightning strike causes a rapid expansion of air in the lightning channel and this explosion of air produces thunder.

CLOUD-TO-CLOUD LIGHTNING
Sheets of cirrus cloud, or anvils, may extend dozens of kilometres beyond the top of a thunderstorm updraft. These highly electrified regions can generate spectacular lightning displays that you can see long before a storm arrives or after it departs.

FORKED LIGHTNING
Cloud-to-ground lightning strikes are stunning to behold, but can be very dangerous to people and animals.

27

Making predictions

Weather forecasting has developed from superstition into science. Yet with a sharp eye, you can spot the basics that drive weather – and make predictions of your own.

WEATHER MAP
Synoptic maps show warm and cold fronts and isobars (lines of equal air pressure).

The professionals

Forecasters predict upcoming weather by feeding observations from across the globe into computers. Highly complex software packages interpret the data, based on our physical understanding of the atmosphere. While still not perfect, one- to three-day forecasts have become far more accurate in recent decades; extended models hint at what weather might arrive as far as ten days in advance.

FORECASTING
Meteorologists draw on data collected daily at weather stations around the globe.

Home weather station

If you're interested in setting up your own weather station, you can choose from a wide range of digital equipment to collect and display daily readings and store them on a home computer. Displays are linked to instruments that measure temperature, humidity, barometric pressure, wind, and precipitation. You can even upload data to public or private networks that collect observations from people all over the world.

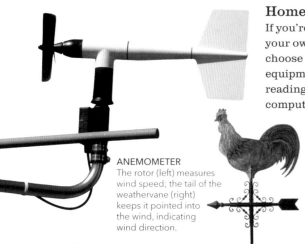

ANEMOMETER
The rotor (left) measures wind speed; the tail of the weathervane (right) keeps it pointed into the wind, indicating wind direction.

DIGITAL WEATHER STATION
By using modern equipment, you can view a detailed portrait of weather conditions, such as temperature and humidity, at a particular location at any time.

air pressure in inches of mercury

FORECASTING A WARM FRONT

In many parts of the world, you can predict a warm front by observing a distinctive sequence of clouds. As a front approaches, warm, moist air sweeps overhead, eroding cold air below. This results in thick, low cloud, and causes an overcast sky for a few hours. Steady rain or snow may eventually develop, ending with a surge of warm air.

1 Thin cirrus clouds (see p.22) are often the first sign of an approaching warm front.

2 A halo or ring may be seen around the Sun or Moon as cirrus clouds thicken into a layer of cirrostratus clouds.

3 The Sun may appear fainter as existing clouds lower and thicken, forming a deck of altostratus clouds.

4 Heavy bursts of rain often occur before a warm front passes, but may end abruptly.

PREDICTING A SHOWER

You can spot an approaching storm by observing cumulus clouds – formed by unstable, rising air (see p.22). When warm, moist cumulus clouds rise into much colder air, they may trigger thunderstorms. Weaker showers grow and die in an hour or two; stronger storms can be set off by a cold front embedded within an area of low pressure (see p.26).

1 Fair-weather cumulus often develop into small, puffy clouds that do little more than block the Sun as they pass by.

2 Moderate cumulus stretch higher into the sky, indicating the risk of a shower or storm in the next few hours.

3 Towering cumulus reach toward frigid atmospheric layers. If they keep growing, a thunderstorm may develop.

Hang seaweed indoors. If it feels moist, it might rain soon.

SEAWEED

Weather folklore

Many cultures have developed unique ways of interpreting Earth's atmospheric behaviour. Common observation threads emerge in folklore surrounding weather. In many different countries, poetic sayings link the look of the sky, or the state of animals or plants, to some future weather event. And while modern forecasting has generally replaced folklore, some of these old weather sayings and practices do, in fact, work.

An open cone indicates warm, dry weather. PINE CONE

RED SUNSET
A red sunset may indicate dry air approaching from the west – hence the saying, "Red sky at night, sailor's/ shepherd's delight."

HALO
Thickening clouds ahead of a warm front produce a halo around the Moon.

HAIL PAD
Sheets of aluminium foil spread on top of polystyrene pads make a simple way of measuring the size of any hailstones that fall in your garden.

RAIN GAUGE
This simple, centuries-old technology remains a standard, reliable method of measuring rainfall.

THERMOMETER
"Max–min" types store the day's high and low readings.

SUPERSTITIONS

Unlike sayings based on atmospheric conclusions, superstitions assign supernatural meanings and explanations to what are really ordinary weather events. One tradition in Britain is that if it rains on 15th July (St Swithin's Day), this supposedly means it will rain for the next 40 days. Rainbows, which are caused when sunlight refracts through moisture in the air, are especially prone to mystical interpretations.

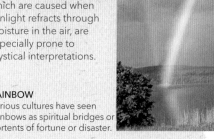

RAINBOW
Various cultures have seen rainbows as spiritual bridges or portents of fortune or disaster.

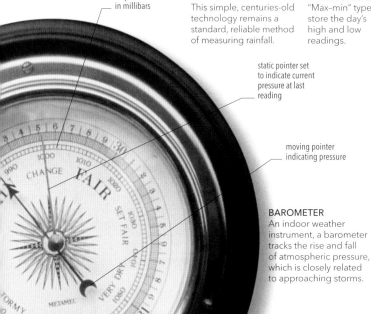

air pressure in millibars

static pointer set to indicate current pressure at last reading

moving pointer indicating pressure

BAROMETER
An indoor weather instrument, a barometer tracks the rise and fall of atmospheric pressure, which is closely related to approaching storms.

Climate change

A global climate has evolved since Earth formed, but now human activity is forcing the atmosphere to change in new and complex ways.

Causes of climate change

Although both natural and human activity have an impact on the evolution of Earth's climate, human causes far outpace the gradual changes produced by various natural causes. Homes, vehicles, factories, and power plants burn vast amounts of coal, oil, and gas, releasing carbon into the air, where it then combines with oxygen to form carbon dioxide (CO_2). This invisible, odourless gas traps heat from Earth in the atmosphere, pushing up the planet's temperature. Airborne CO_2 has increased 35 per cent since the 1950s and global temperatures keep rising – with potentially dramatic results for habitats and wildlife.

NATURAL CAUSES
Major volcanic eruptions actually cool Earth's climate by releasing sulphur dioxide, which reflects sunlight, but they also add carbon dioxide to the atmosphere.

A big eruption throws dust and gases into the atmosphere.

HUMAN CAUSES
Burning coal for power is one of the most significant human-produced causes of carbon dioxide in the atmosphere.

PLANTS AND CLIMATE CHANGE

In warm regions, trees help cool the climate by shading the soil and trapping moisture. Planting trees in these areas may help offset global warming. Plants also absorb carbon, but the carbon eventually returns to the air whenever a plant dies and decays. Conversely, trees have a warming effect in subpolar regions, where dark evergreens absorb more sunlight than snow-covered ground. Plants and oceans absorb about half the CO_2 emitted from human activity, and plant-stressing droughts can have the effect of cutting that absorption in half for up to a year.

EARLY SIGNS OF SPRING
In England, where the timing of the first bluebell flowers has been tracked for decades, the blooms now arrive about six days earlier than they did in the 1950s.

Signs of change

Many natural indicators point to a warming climate. Most glaciers around the world have retreated in the last century and the Arctic Ocean is losing more sea-ice in summer. As air warms, more water evaporates into it, so drought-stricken ground tends to dry further while heavy rains become heavier. Sea levels are rising as glaciers melt and oceans warm and expand. Climate change threatens the habitats of many plant and animal species, which could lead to their extinction.

RISING SEAS
The average sea level has risen nearly 20cm in the last century. Coral atolls and low-lying deltas are most threatened by further rises.

GLACIAL THAW
Polar bears and other creatures rely on the Arctic Ocean's sea-ice. The average extent of late-summer ice has dropped more than 40 per cent since 1980.

Tracking changes

Amateur naturalists around the world help scientists keep track of changes in plants, animals, and insects as the climate warms. Through programmes such as Project BudBurst (see panel, right), and the National Phenology Network in the USA, Nature's Calendar in the UK, and other similar projects worldwide, volunteers record changes they see and report their observations on the internet. These efforts will become more and more valuable over time as climate change unfolds and the amount of data grows.

PROJECT BUDBURST

Thousands of Americans take note of spring's progress each year through Project BudBurst, a programme sponsored by a museum, a university, and a research centre. The volunteers record such events as buds opening and fruit ripening. Hundreds of plant species are now being tracked through the project, and thousands of observations are submitted each year.

CYCLES OF CHANGE
Once endangered in the UK, the comma butterfly is now more frequently observed. Its adaptable life cycle has allowed it to expand its range as far north as Scotland, responding to favourable weather conditions.

Night watch

Nothing stretches the imagination and prompts us to question our place in the universe like stars. The sky on a crystal-clear night is an incredible sight, and it is surprising how much you can see.

Use binoculars to see even more stars.

The Moon

The Moon is a wonderful, even magical, sight. It is our nearest neighbour in space and revolves in time with its orbits around Earth, which means we only ever see one side of it. It is best studied when the low, raking light of the Sun picks out its mountains and crater walls in sharp definition. A full Moon reflects brilliant, intense light and minimizes contrast so, although you can see the entire Moon, it is the least rewarding Moon phase for observation. The brilliance of a new Moon also tends to overwhelm nearby stars.

Seeing stars

Stars are grouped into areas of the night sky called constellations, with names such as the Southern Cross and Orion. The stars in a constellation have no connection to each other – the "patterns" are created by chance and early civilizations named them partly to help with navigation and orientation. Knowing where to find constellations will help you understand the night sky better. Apps on phones or tablets can also help identify and locate them. Binoculars or a small telescope will help you to see more stars than the naked eye, and to enjoy the constellations and star clusters to the full.

MOON LIGHT
Moon phases are caused by the Sun lighting up the Moon. A new Moon is dark because the Sun is lighting up the side furthest from the Earth. As the Moon orbits Earth, more can be seen until it is all visible at a full Moon.

STAR MAP
A star map is essential for navigating the night sky. Pick out the Plough or Big Dipper, shown on this map.

MEASUREMENTS

The size of celestial objects and the distance between them are described in degrees and parts of a degree. Calculating degrees can be done by simply using your hand as a ruler. Hold it up to the sky, at arm's length, and use these standard measurements to help make your calculations.

FINGERTIP DEGREE
A finger width, held at arm's length, measures about 1° across.

JOINT MULTIPLES
Finger joints are roughly 3°, 4°, and 6° across.

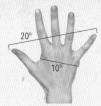

HAND AND PALM SPAN
An average adult handspan covers 20°; the palm is 10° across.

PREDICTED SIGHTINGS

Stars are fixed objects, but planets and other bodies move around the sky. Astronomers can predict where most will be on a given night, and many resources can tell you what to look for each month, such as which planets can be seen, or whether meteor showers or a bright comet are due.

COMETS

METEORITE

Planets

You can see Venus, Jupiter, Saturn, Mars, and Mercury with the naked eye or with binoculars. Venus and Mercury are evening or morning objects. Venus is usually the brightest of the planets, and all of them, except Mercury, are brighter than any star. Uranus and Neptune are faint compared to Venus, and you need binoculars, a precise location, and a star chart to spot them. Unlike stars, which are very far away and appear as points of twinkling light, planets are solid, much nearer to Earth, and do not twinkle.

MORNING LIGHTS
Venus and the Moon at dawn make a spectacular sight. At its brightest, Venus is more striking than any star. It is so bright that it has even been mistaken for an alien airship or UFO.

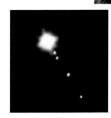

THROUGH BINOCULARS
Jupiter is so large that binoculars show its defined shape. Watch it on different days to see its brightest moons move around it.

Using telescopes

You can get great results from a telescope without spending a fortune. Choose a model with a wide lens or mirror – high magnification is far less important – which will gather maximum light and help you to see objects such as faint stars, clusters, nebulae, and galaxies. A "go-to" system automatically points some telescopes to your chosen target and is often affordable. Balance cost with practicality, and always visit a good dealer for advice.

lens

eyepiece

counterweight

equatorial mount

leg and tripod

EQUATORIAL MOUNT

ALTAZIMUTH MOUNT

FOLLOW THE STARS
Mounts help telescopes follow the stars. An equatorial mount needs to be carefully aligned, while an altazimuth – such as the Dobsonian altazimuth mount shown here – moves freely from side to side and up and down.

What a naturalist needs

In these gadget-obsessed days, the excitement of preparing a "tool kit" that contains all the essentials for properly exploring the natural world is no longer the sole preserve of the geek. Whether young or old, being technically prepared is part of our lives, and there are plenty of new toys available for the modern naturalist. Field guides for mobile phones, tiny cameras to reveal the private lives of nestlings, and chemical lures for specific moths join affordable night-vision binoculars and bat detectors that transfer recordings to your PC as off-the-shelf goodies. But remember, the most critical part of the kit cannot be bought. It's a lifetime of curiosity!

A naturalist's kit

Curiosity, enthusiasm, and common sense are some of a naturalist's most important tools. Add a field guide and some way of taking notes, and you are well on your way.

Observing and recording

Becoming a naturalist is about becoming a systematic observer of nature. To do this, you need to have some way of recording what you have seen. It is only when you begin to log your discoveries that patterns and trends begin to emerge – which in turn make you a more focused observer. Some field guides (whether in printed form or digital) for identifying species and a notebook and pencil for jotting it all down are all you need to get started, but don't forget all the time-saving devices offered by modern technology.

DIGITAL NOTES
Modern mobile phones provide numerous ways of taking on-the-spot notes in the wild. You can take quick snaps or save observations as text messages or voice recordings. Use your phone's digital camera to create a high-resolution record of your observations during the day. Most phones will add date and location information to the image files.

FIELD GUIDES
Either carry field guides with you, or take detailed notes of species you are not familiar with to identify later.

NOTEPADS, PEN, AND PENCIL

ADHESIVE TAPE

KEEPING A RECORD
Writing your observations in a notebook is an easy and economical way to keep a lasting record. Sketch your impressions and describe species in detail. Use tape to attach findings such as feathers and leaves to the pages.

ACCURATE NOTES
When making notes, either digitally or on paper, try to be as precise as you can to make comparisons easier later on. Always remember to record the location and date of your sightings. You may want to carry a ruler or tape measure to accurately record details such as size.

TAPE MEASURE

RULER

SPECIALIST EQUIPMENT

Some animals are hard to observe well. Bats, since they fly are night, are almost impossible to identify on dark nights. However, a bat detector's ultrasound microphone converts their ultrasonic echolocation calls to frequencies that humans can hear. The animals can then be identified by the rhythm and frequency of these calls. Many moths also fly unseen at night, but they are attracted to light-traps, where they can be observed closely. Birds may not allow close approach, and sometimes even binoculars are not powerful enough to give good views. But birding telescopes that magnify 70x or more can overcome this problem.

BAT DETECTOR
A hand-held bat detector detects the ultrasonic calls of bats and converts them to the human hearing range.

ECHO METER
A device with an ultrasonic microphone can be attached to a phone or tablet so that you can record and identify bats.

wide lens catches light

TELESCOPE
For steady close-ups use a scope on a tripod. An angled eyepiece makes viewing more comfortable.

MOTH TRAP
Record the moths that live in your garden with a lightweight, battery operated light trap.

USING BINOCULARS

Binoculars allow you to get close to wildlife while causing the minimum amount of disruption. To balance any difference between the two eyes, first cover the right side; use the central wheel to focus a sharp object – like a TV aerial – with your left eye. Now cover the left side; use the right eyepiece adjustment to get the same object sharp with your right eye. Look straight at what you want to see, then bring the binoculars up to your eyes. Use only the central wheel to focus on different distances.

central focusing wheel

eyepiece adjustment

A STEADY GAZE
Use both your thumbs and fingers to help keep binoculars steady.

CLOSE-FOCUS BINOCULARS
Some compact binoculars allow you to focus on subjects at close range. These are ideal for viewing butterflies and dragonflies.

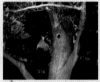

NAKED EYE VIEW

8x MAGNIFICATION

20x MAGNIFICATION

MAGNIFICATION
A steady view magnified 7 or 10 times shows more than a wobbly one twice as large. Choose binoculars with a magnification between 7x and 14x at most.

A closer look

In order to learn more about certain plant and animal species, it is useful to take a closer look. For some difficult groups, a magnifying lens is essential for accurate identification. Try to observe the specimen as you find it and never collect wild plants or animals. If you catch an insect for closer study, always handle it with great care, and release it afterwards.

EASIER VIEWING
Clear perspex boxes are best for observing insects and other small animals more closely, allowing you to view the specimen from all sides with minimal disruption. Some containers have a built-in magnifying lens for easier viewing.

COLLECTING CONTAINERS

LOOKING AT DETAIL
Use either a loupe lens or a larger magnifying glass to record details such as whether a flower stem is smooth or hairy, the shape of a beetle's jaw, or the wing structure of a dragonfly. Tweezers are useful for holding up small specimen you may find; live animals should only be observed in boxes.

CLOTH

TAKING SAMPLES
Never uproot a wild plant – if in doubt, leave well alone. Use a knife to take a leaf sample for identification later on.

LOUPE LENS

TWEEZERS

MAGNIFYING GLASS

CATCHING INSECTS
In order to catch insects for closer observation, make sure you have the right equipment such as a net or pooter. Take great care not to damage the animal in any way and always release it afterwards.

NET

POOTER

BUTTERFLY NET

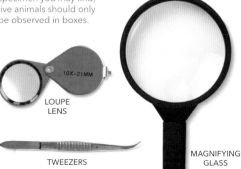

Being prepared

Life outdoors is unpredictable, but a changeable forecast is no excuse to stay in. You can take simple steps to be comfortable in most weather conditions.

Evaluating conditions

For the most part, the time of year, the weather forecast, a good map, and an idea of exactly what you are trying to achieve prepares you for most eventualities outdoors. Dress and pack accordingly, bearing in mind that you may be carrying everything with you all day. If buying new equipment or clothing, choose comfortable, lightweight options that allow maximum ease of movement. The most useful gear doesn't have to be expensive – your aim is to avoid being too cold, hot, wet, or getting sunburn.

HEAD TORCH

HEADGEAR
Always remember to protect your head outdoors. A warm hat is invaluable on a cold day, while a cap and sunglasses provide shade from the Sun.

HAT

CAP

POLARIZING SUNGLASSES

LAYERING
Wearing several thin layers allows you to easily add or remove clothing as conditions change. Layering is also the best way to keep warm – the air trapped between the layers is an efficient insulator.

fingerless gloves keep hands warm while writing or using equipment

FINGERLESS GLOVES

fleece jacket combines comfort with warmth

lightweight jacket is easy to wear and breathable, yet keeps you dry

base layer wicks perspiration away from body

waterproof zips ensure gear stays dry in all conditions

external pockets provide quick access to gear

TROUSERS

Choose trousers that are both robust enough to protect you from scratches and lightweight enough to be comfortable. Trousers with zip-off legs are ideal for changeable weather or an impromptu paddle.

zip-off legs adjust for warm weather

RUBBER BOOTS

HIKING BOOTS

SANDALS

FOOTWEAR

Choose footwear according to the terrain. Sturdy sandals can be worn on flat ground, while hiking boots are ideal for rough terrain. Rubber boots can be useful when the ground is wet or marshy.

THE DAY PACK

The best rucksacks sit high on the back for comfort, have an abundance of zipped pockets, and are waterproof to keep out the rain. Like all outdoor gear, try it on before you buy. A belt bag is the ideal choice for short treks.

waist strap helps distribute weight evenly, sparing shoulders

BELT BAG

WET GEAR

What you need depends on what you're doing, when, and where. A swimsuit, mask, and snorkel are all you need to explore aquatic wildlife when the weather is warm. A wetsuit is a good investment for snorkelling in cooler climes, or if you plan to spend long periods in the water.

MASK AND SNORKEL

FLIPPERS

SWIMMING COSTUME

Getting around and staying safe

If you're heading further afield than your local area, make sure you've planned your excursion and have the right equipment to hand.

Finding your way

Exploring a new area requires a bit of preparation and research, but need not be difficult. A map or trail guide is an essential piece of kit – simple ones are often available online. Plan your route in advance and have a map to hand even when following marked trails to ensure you make it back before nightfall.

USING A COMPASS

A compass is useful whenever you are using a map, but essential if walking on land outside marked trails. A compass, at its simplest, is an instrument that always points to the magnetic north. Use it to ensure that your map is orientated correctly, and to check your course of travel as you walk from point to point.

FINDING NORTH
Align your compass with map gridlines and rotate both until the needle points to "N".

direction-of-travel arrow   orientation arrow   magnetic needle

MAPS
A wide variety of maps is available for navigating in wilderness areas. Choose a large-scale contour map for maximum detail about the various features of the landscape.

TORCH
With no other lights around, darkness can fall very quickly. Make sure you always carry a working torch.

WATCH
Keep track of the time to check your progress and assess when you need to turn back.

GPS DEVICE
Smart phones use GPS to calculate your position and show it on a map.

PUBLIC RIGHTS OF ACCESS

Rights of access refer to the rights of the general public to use public and privately owned land for recreation. Access rights can vary considerably from country to country so always obtain up-to-date information from local authorities when visiting a new area. Access rights can be limited to rights of way, meaning that access to land is only permitted via a certain path or trail.

RIGHTS AND RESPONSIBILITIES
Rights of way are there to be used, but remember to close gates, avoid leaving litter, leave plants alone, and, if you have one, clean up after your dog.

The Countryside Code

1. Plan ahead, be safe, and follow any signs.

2. Leave gates and property exactly as you found them.

3. Protect plants and animals and take your litter home.

4. Keep dogs on a lead.

5. Consider other people.

NAVIGATING IN THE WILD
It takes some time to learn to use a map and compass confidently, but it is a skill that will pay off time and again.

Staying safe outdoors

There are a few things to bear in mind to ensure your comfort and safety outdoors. Watch regional weather forecasts and be prepared – good conditions can turn bad quickly, especially on exposed ground and mountains. Be careful near the sea, especially on beaches and salt marshes with a large tidal range – you can find yourself a long way from safety when tides turn. Use a tide table (see p.180) and watch out for strong winds. Though livestock rarely present problems, it is wiser to avoid them if you can. Take precautions in unfamiliar places on your own, or as it gets dark. And always tell people where you're going and when you expect to return.

BITES AND STINGS
Spiders and insects rarely cause problems, but find out about any dangerous ones in the area.

INEDIBLE PLANTS AND FUNGI
The golden rule is, simply, don't eat anything, unless you are absolutely sure of what it is.

SEVERE WEATHER
Don't venture far afield in difficult terrain if the weather looks bad – and always take suitable clothing.

EMERGENCY KIT

A mobile phone is a safety essential, but you may not have coverage everywhere. Remember to programme it with emergency numbers. Always pack a whistle – it can help rescue teams find you if you're unable to move. Be sure to also pack water, high-energy snacks, and basic first-aid supplies, especially if you're planning a long hike in unknown terrain.

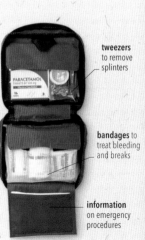

tweezers to remove splinters

bandages to treat bleeding and breaks

information on emergency procedures

FIRST-AID KIT

WHISTLE

Photography

Taking photographs is a great way to record the natural world. Using the correct equipment and techniques can help you to get some amazing shots.

Choosing your camera

It is the opportunistic and unpredictable nature of wildlife photography that makes capturing a great image so rewarding. Expensive gear is all very well, but being in the right place at the right time, and being alert, patient, and respectful is key. Regardless of what you want to do, you should be familiar with your camera, and it should be comfortable to use and kept with you at all times.

Get as close as you can to your subject without disturbing it.

shutter — shutter dial — flash

BRIDGE CAMERA
Relatively small and light, a bridge camera enables switching from wide-angle to telephoto mode with the movement of a dial rather than the inconvenience of changing lenses.

shutter

changeable lens

DIGITAL SINGLE LENS REFLEX (DSLR) CAMERA
The camera of choice for serious photographers is the DSLR. These cameras have a range of features that can be adjusted manually and also offer greater flexibility with interchangeable lenses.

CLIP-ON LENS
A smartphone clip-on macro lens allows you to get close to the subject and pick up fine detail. Attached to a phone, it is ideal for insect and flower photography.

lens attachment

CLIP-ON LENS UNATTACHED

sturdy clip secures around the smartphone's camera

CLIP-ON LENS ATTACHED

USEFUL EQUIPMENT

A tripod is useful when your camera needs to be stable, whether for close-ups of tiny plants or insects, or to steady your hand when using a zoom lens. Your kit should also include a lens-cleaning cloth, blower brushes, spare batteries, and digital cards.

strap to attach tripod

camera is protected by padding

CAMERA BAG
Expensive equipment needs protection when on the move. Carrying bags offer sturdiness, waterproofing, and balance with weight and comfort.

ZOOM LENS

WIDE ANGLE LENS

EXTRA LENSES
Zoom lenses are most suitable for close-ups, while wide angle lenses are ideal for landscape photography.

Photographing wild animals

Animals in the wild are fickle subjects due to their unpredictable natures – you must take care not to approach or disturb potentially dangerous animals. Studying the techniques used by professional wildlife photographers can improve your chances of getting great shots at a safe distance. "Digiscoping", for example, where you fix a camera to a telescope via an adapter, allows you to shoot close-ups without disturbing your subjects or their environment.

HIDES

A hide, or blind, is a shelter that offers protection for you and your equipment, and gets you close to your subject without being seen. Fixed shelters, such as a car, can be effective, but a hide built over a period of days to avoid alarming the subject, is often better. Tents can be adjusted for this purpose.

Use a waterproof camera to photograph underwater life.

Light and exposure

To capture wildlife in action, and get a good depth of focus, you usually need as much light as possible for quick exposures. But don't look just for bright sun "spotlighting" the subject; think about different types of light, the moods they create, and the effect you want. Go out both early and late in the day and observe the play of light and shadow. Try taking shots with low, raking sunshine, or with light reflected by mist or snow. And don't forget the flash – it is essential for images in low light conditions.

1 A pheasant captured at full speed. Bright light allows a fast shutter speed to catch the action. A slow shutter speed would create a blur.

2 Bright, autumn sunlight has enriched this pheasant's colours. Dull light flattens colour, but it can create a more subtle, atmospheric result.

3 An owl at night taken with a flash, probably triggered by an electronic motion sensor.

4 A subject, like a bird, can be emphasized by isolating it from its background using a shallow depth of field. Create this effect by using a large aperture.

Keeping a record

Recording your observations not only helps you learn more about what you have seen, but lets you contribute to the data-collecting efforts of the naturalist community.

Field notes

Whether digital or hand-written, the notes you take in the field form an important part of being a naturalist. Note-taking helps focus your attention on the details of what you see – instead of a brown bird, you will learn to see a smallish brown and cream bird, with dark brown markings. Take photos or make sketches to form a more comprehensive picture, and use your notes to look up and learn the names of new species later on. A log of repeated observations allows you to link your discoveries to wider natural phenomena.

PHONE APPS

Apps designed for phones are great tools for naturalists. Apps are available that identify birds from pictures you take, and sounds that you record, with your phone. There are also apps that identify bats by their echolocation calls and plants from photographs.

Bolt-on ultrasound microphone device

Birdsong audio waveform

remember to note date and location of sightings

always include name and any interesting information

note colours of plants and animals

make sketches and add details for later identification

include details of environment in which each species was seen

12/10/2010
Location Ham Lands
Nature Reserve,
Richmond

– Rowan berries just changing colour from orange to red – earlier than last year!

– Berries seem especially plentiful this year; due to wet summer?

– Grey-brown caps and light-brown stems
– Grows in clumps of 2–3

Growing in the shade of trees, almost hidden by fallen leaves

Stems 10–15cm in length, caps up to 5cm across

pale underside

Collared dove feather – under tall trees

SKETCHING

If you look at something superficially you will soon forget most of the details. If you take time to observe it and draw sketches, however, it will stick in your mind for years. You don't need to be a great artist to sketch the basic shape of an animal such as a bird. Record the overall shape first, then add the details such as tail shape and markings.

1 Use basic ovals to create a bird shape. Get the rough shapes and proportions down, then add a basic bill, tail, and legs.

2 Fill in the general shape and revise the bird's proportions. Is it upright, horizontal, slim, chubby, or long- or short-legged?

3 Add essential details: shapes, relative proportions – do the wings reach halfway along the tail or fall short? – and feather patterns.

4 Look closely at the head and note down the colours and pattern of the plumage and the shape and relative length of the bill.

5 Make sketches from different angles, if possible. Have you labelled everything? If not, take another look. Making notes around a sketch forces you to notice specific characteristics.

Data collection

If taking notes on a regular basis, it is a good idea to transfer the data you collect in the field into a more systematic form for easier access later. Use field guides and apps to identify species and compare your observations with other people's. Don't forget to add or scan in your photos and sketches. Regular, detailed records of the same subjects can help build up a set of valuable data not just for yourself but for the wider naturalist community. A single count of birds on a lake is interesting, but a series of weekly counts taken for a month or year can have real scientific value. Look for local or national surveys to which you can submit your findings. You can use a phone app to enter your observations to an online database such as eBird or iNaturalist. By doing so, you are contributing as a "citizen scientist" to global knowledge of species' status and changing fortunes – and you can refer back to your previous records as well.

note any items collected along the way

ORGANIZED SYSTEM
Uploading your records onto your mobile phone or computer allows you to easily retrieve data collected over time. Use online apps to share your records with the wider naturalist community.

BIRDWATCHING
Your data can form part of a regional or national bird survey. Programmes all over the world, such as www.ebird. org in the US, pool together data from naturalists.

Close to home

Tropical rainforests and Antarctic seas are home to many celebrity species, but it is in and around our own homes that we meet most of our wildlife. These encounters are formative at first, then ultimately much more rewarding than fantasies of exotic creatures fuelled by television. To watch in "real time", to identify species that share our community, to relate to their lives and perhaps provide for them – even touch, feel, and smell them – this is the essential source of a real affinity with nature. And despite our worst efforts, so many animals and plants have managed to adapt to sharing the **human environment** that meeting them can be an everyday event.

Home

We all have a number of visitors in our houses. Wasps, birds, and bats may nest in your loft or roof space, while beetles and termites might burrow into wood. Peer into cracks in walls and you may find mice or cockroaches. Look around to see what's sharing your space, but don't view anything as a "pest" – these animals don't exist to aggravate us, but as part of the complex system of nature.

COMMON WASP

MEDITERRANEAN HORSESHOE BATS

Local habitats

You don't have to venture far to explore the natural world. Our homes, gardens, parks, streets, and railways are teeming with wildlife – if you know how to look for it. Many animals and plants can live alongside humans.

Garden

Gardens are great for watching wildlife, especially if sensitively managed with nature in mind. Even the smallest outdoor space – a window box, terrace, or patio – can be stocked with plants to attract insects and the animals that feed on them. In larger gardens, ponds, compost heaps, log piles, and wilderness areas provide more opportunities for animals and plants to thrive. Bird baths, feeding stations, and nest boxes attract wildlife, too.

COLLARED DOVE

COMMON DAISY

GARDEN SNAIL

Town park

Urban parks are oases of green that provide a much-needed "breathing space", both for people and wildlife. Established parks have mature trees and may include a pond, lake, and wildlife area. Look for songbirds, waterfowl, bats, and other small mammals such as squirrels, and hedgehogs.

SWEET CHESTNUT

GREY SQUIRREL

Street

You may think there is little space for wildlife in the concrete jungle. However, look closely and you will see wildflowers growing through cracks in the pavement, and insects burrowing into the mortar of a wall. You might even catch sight of a rat or a fox scavenging in dustbins, and don't forget to look up to spot birds roosting on buildings and street lamps.

BROWN RAT

ZINNIA

Railway

Next time you take a ride on the train, look out of the window for wildlife. Some plants, including ragwort (a type of daisy), flourish in the well-drained, gravelly conditions along tracks, and you may see mammals, such as wild deer. Disused railways are often converted into footpaths, while derelict tunnels can be adapted for hibernating bats – only visit such areas with caution.

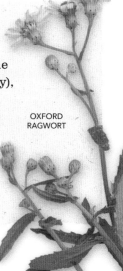

OXFORD RAGWORT

In the home

Our homes provide shelter for animals other than ourselves. We share our living space with an array of successful opportunists.

House guests

The most successful species in nature are those that can quickly adapt to change and capitalize on opportunities as they arise. As humans alter the landscape wildlife must adapt to survive. Some animals have evolved to find their niche in our homes, which meet two basic needs – shelter and food. For example, a dry attic is a perfect place for a wasp to build her nest, while discarded food waste is a feast for a house mouse. Each species has its place in the world and many, such as spiders, provide a valuable service by keeping the levels of other house guests, such as flies, in check.

LIQUID DIET
Houseflies can contaminate food by passing bacteria from their feet and mouthparts. They suck in liquid food through a fleshy proboscis.

TAKING A BATH
Some spiders can bite if threatened, but they come inside to feed on insects, so don't harm them.

AN ITCHY VISITOR
Adult fleas need blood in order to reproduce. If you notice small, irritating bites, usually on legs and ankles, there may be fleas in your house, if you have mammalian pets. Their eggs often drop into bedding or carpets, where the larvae feed and pupate. New adults then jump on to a passing host.

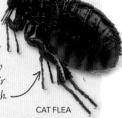

Fleas have long back legs and can jump up to 350 times their body length.

CAT FLEA

GECKOS
Geckos make good house guests. They keep insects, such as mosquitoes, in check.

LOFTY AMBITIONS
Bats often roost in loft and roof spaces. They will not gnaw on wood and do no structural damage – their dry droppings are also rarely a nuisance.

Signs of life

Some visitors are unseen until you notice their tell-tale signs. If you live in the USA, Africa, or Australia, you may not be aware of a termite invasion until you see a nearby swarm, but they could have already been at work in your home. Other visitors are more obvious, and you'll see droppings or nests. You may also hear scratching or chattering in a ceiling or wall, or smell a distinctive odour.

LITTLE NICHES
Small holes in your furniture can be a sign of "woodworm": the common name for the beetle larvae that bore into wood.

HOLES IN CLOTHES
If you see small, light brown moths in your home, check your clothes and carpets. Adult moths do not eat, but their larvae feed on natural fibres, such as wool.

BLISTERING PAINT
Buckling paint may mean termites are around. They can cause damage to structures by eating through wooden supports.

TINY DROPPINGS
Small, black pellets in cupboards or on floors are often a sign that mice are inhabiting your home.

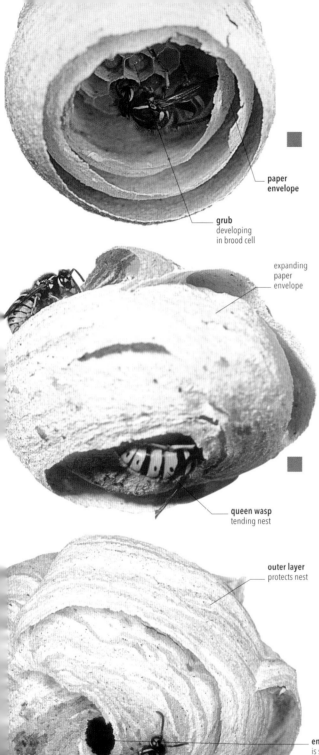

paper
envelope

grub
developing
in brood cell

expanding
paper
envelope

queen wasp
tending nest

outer layer
protects nest

entrance hole
is small and
easy to defend

worker wasp
continues to
enlarge nest

Homes from homes

A nest is usually an indication that an animal has set up home in your house. Wasps' nests are common in attics, garages, and lofts – try to look at them from a distance, or with binoculars. Basements, garages, and roof spaces are good places to hunt for the nests of small mammals, such as mice, while evidence of nests around your house may indicate that birds are living in your roof. Swallows and martins build their nests outside under roof overhangs, so you can watch them as they work to feed their young. Never disturb a nest unless it is unavoidable.

LONE QUEEN
A queen wasp uses her antennae to tend her nest.

Wasps' nest

1 Social wasps live in colonies in nests. A solitary queen begins the nest, laying a single egg in each brood cell as she completes it. The nest is a sequence of paper layers, made out of chewed wood fibres.

2 The queen tends and feeds her growing grubs with the caterpillars she has caught until they hatch into worker wasps. The workers then help the queen to expand the nest, allowing her to spend more time producing eggs.

3 The nest has a small entrance hole that is easy to defend and also helps to control the interior temperature and humidity. Workers continue constructing new outer envelopes to accommodate the growing colony.

HOUSE MICE
Mice build nests out of materials they find in and around homes, such as dry grass and soft fabric fibres.

Swallows feed their young on insects caught in flight.

MOUTHS TO FEED
Look up, you may see birds such as swallows and martins building their nests beneath the eaves of houses or outbuildings.

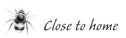

Spiders

HOUSE
SPIDER

Spiders are common in homes and gardens. The best way to find them is to look for their silken webs.

Around 45,000 species of spider have been recorded worldwide. They build their webs out of lines of silk, expelled through silk-spinning glands (spinnerets) at the rear end of their opisthosoma (abdomen). Some spiders maintain and repair the same web for some time, while others eat their webs in the evening and construct a new one the next day. The intricate webs of garden spiders are a spectacular sight on a dewy morning. Search carefully among bushes and shrubs, but be careful not to touch any part of the web or the spider will hide. Plain, brown house spiders do not construct such beautiful webs – look for flat, tangled webs in parts of the house that are not used very often. Other spiders you might see around your garden or home include wolf spiders, the females of which often carry a white egg sac under their abdomens; jumping spiders, such as zebra spiders that stalk their insect prey; and tiny money spiders, some of which construct fine, sheet-like webs in grass and hedgerows.

BANDED
GARDEN
SPIDER

TYPES OF WEB

Different spiders spin different types of web. Some spiders create a radial web with strands of silk extending around it that act as tripwires to alert the spider when an insect touches them. House spiders weave a tangled sheet of silk to catch insects that crawl or fly into it, and garden spiders construct orb webs across gaps to catch flying insects.

RADIAL WEB

SHEET WEB

ORB WEB

CLEVER CONSTRUCTION
Most orb web spiders, such as this European wasp spider, take less than two hours to spin their webs. They use the non-sticky rays (spokes) of the web for transit.

Round the clock

Garden visitors change over a 24-hour period. Diurnal animals are active in daylight hours, while those awake at night are nocturnal.

Daytime

Go out into the garden at dawn, as the first rays of sun peek above the horizon, and you will hear songbirds begin their dawn chorus (see p.98). As the Sun gathers strength, butterflies, dragonflies, and reptiles come out to bask in its warmth. Garden birds start to appear as the day wears on, you may notice them busily feeding themselves and looking for food for their young.

FIND A FOX
Foxes can be seen both during the day and at night, a pattern also followed by rabbits.

FOLLOWING THE SUN

Plants that turn to follow the Sun as it moves across the sky during the day are heliotropic, and the motion is known as heliotropism. They track the Sun with their leaves to maximize the amount of light they absorb for use in photosynthesis. Sunflower blooms also follow the Sun to attract insects that favour its warmth.

NIGHT OR DAY?

Animal behaviour is influenced by a biological process called circadian rhythm, a daily cycle that provides cues as to when to sleep, wake, and feed. Many animals (including humans) are diurnal, and remain active during the day, but others are crepuscular (active during twilight hours), or nocturnal (active at night, see p.14). Nocturnal animals may have adapted this behaviour to avoid competition for food with similar diurnal species, to avoid dehydration in hot habitats, or to avoid predation.

DAYLIGHT HUNTER
Kestrels are diurnal birds of prey. As they can hover for extended periods, they can survive in a variety of habitats, including city centres. Look for them hunting by major roads.

SPOT A SQUIRREL
Squirrels are busy during the daytime, feeding and storing food for later. They are agile and very good at stealing food from garden bird feeders.

SEE A SNAKE
Reptiles, such as snakes, bask in the morning sun to warm their bodies and speed up their metabolism.

LET FRUIT LIE
Leave fallen fruit on the ground in your garden to attract daytime feeders, such as birds and insects.

Night-time

If you sit quietly in your garden at dusk, you might see bats. They fly and hunt insects in the dark by using sound waves to create a mental image of their surroundings. This is called echolocation. Other nocturnal creatures often have a heightened sense of sight, smell, or hearing. Torchlight reflected in the eyes of a fox or cat is due to a special layer in the retina, called the tapetum, which takes in the maximum amount of light and gives them excellent night vision.

RUBBISH RAIDERS
Some mammals, such as red foxes, will scavenge in rubbish bins at night and by day.

MAKE A SAND TRAP

A sand trap can tell you which animals visit your garden after dark. Spread a thin layer of sand in an area you think they may use regularly, perhaps a run under a garden hedge or around a feeding station, where you can leave some pet food to attract them. In the morning, see what prints have been made by visiting creatures and try to identify them.

mouse footprint

SANDY PRINTS
Various animals might be tempted to cross the sand. These prints belong to a hungry mouse.

TEMPT A MOTH
Plant scented, night-blooming flowers to attract moths. The scent helps them to find the flowers in the dark.

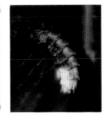

GO FOR GLOW
Tiny lights in rough grass at night indicate the presence of glow worms, a type of wingless, female beetle or larvae.

FEED A BADGER
If badgers are in the vicinity, you might attract them into your garden with peanuts, pet food, and meat.

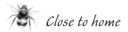
Garden birds

Watching birds in the garden can be truly rewarding. You can encourage your avian visitors by providing food, shelter, and water.

Attracting birds to your garden

The more habitats your garden provides, the better it is for birds and the more species you will attract. Provide shrub and tree cover for shelter and nesting places. Choose native plants that produce seeds and berries as well as those that attract insects, and avoid highly invasive species such as buckthorn. Offer a range of food on bird tables and in feeders and clean water for drinking and bathing, and put up nesting boxes (see p.61).

AUTUMN BERRIES
Hawthorn fruits in autumn are a favourite of berry-eating birds such as this mistle thrush, as are the berries of related species like cotoneaster.

SUNFLOWER
You can buy sunflower seeds for your bird feeders, but why not plant the flowers? They not only attract insects, but once they have bloomed you can keep the dried heads for birds to feast on the oil-rich seeds.

BLACKTHORN

SAFE NESTING
Holly bushes provide safe nesting and hiding places for small birds, such as robins. The bright-red berries are eaten by larger birds, including mistle thrushes, fieldfares, and blackbirds.

Spotting the birds

Enhance your observations by attracting birds to a feeding station or bird bath. Watch quietly from a window and don't make sudden movements that might startle your visitors. Depending on your location, you may spot birds from these common groups.

Thrushes and chats

Thrushes and chats are small to medium-sized birds renowned for their beautiful songs. Most thrushes are brown or grey, with speckled underparts – but some of the closely related chats are brightly coloured. Thrushes eat insects and berries. Some also eat snails. Watch a song thrush use a stone to break a snail shell.

Starling family

Starlings usually have dark plumage with a metallic sheen. European starlings have white speckles, especially in winter, and can form huge flocks to feed and roost (see pp.72–73). They have been introduced to the Americas, Australia, and New Zealand. Other species of starling include Asian myna birds and African starlings.

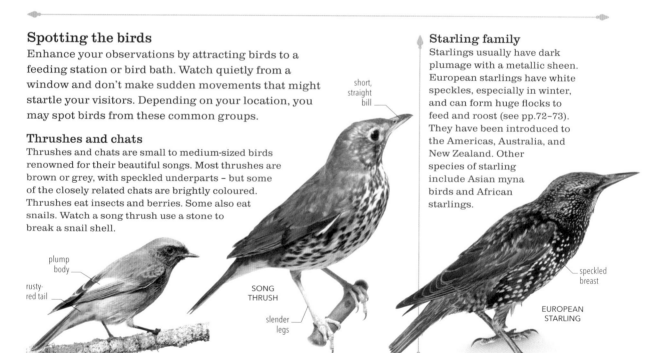

short, straight bill

plump body

rusty-red tail

SONG THRUSH

slender legs

BLACK REDSTART

speckled breast

EUROPEAN STARLING

Woodpecker family

Woodpeckers are boldly patterned birds ranging in size from small at about 15–18cm long with the largest being about the size of a crow. They have sharp, chisel-shaped bills for pecking or hammering holes in trees to find grubs and make their nests. Two of their toes face forwards and two backwards to help them grip.

red crown

green back

stiff tail for support

GREAT SPOTTED WOODPECKER

GREEN WOODPECKER

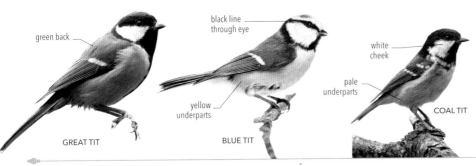

green back

black line through eye

white cheek

yellow underparts

pale underparts

COAL TIT

GREAT TIT

BLUE TIT

Tit family

Birds of the tit family have small, round bodies and short, triangular bills. Some have crests, and some have very long tails. Many are brightly coloured. North American members of the family are called chickadees or titmice.

Finch family

The finches are small birds with many plumage colours, including browns, greens, and blacks and some striking reds and yellows. Their bills vary widely in size and shape according to their diet. In addition to bullfinches, goldfinches, and greenfinches, the family contains grosbeaks, linnets, and siskins.

EUROPEAN GREENFINCH

forked tail

Crow family

Crows are medium to large birds with strong, scaly feet and stout, robust bills. Many species are black or grey, but some are pied – for example, the magpie is white and black. Others, including Eurasian and Siberian jays, are brightly coloured. The family also includes jackdaws, rooks, and ravens.

blue panel in wing

CARRION CROW

strong feet

EURASIAN JAY

BIRDS IN FLIGHT

Identifying flying birds is often challenging. However, a few clues may help you work out the family at least. Besides the rhythm of the flight, things to note include the size of the bird (often difficult to gauge at a distance), its wing shape, tail shape, colour, and how it holds its neck and legs. For example, herons fly with their necks bent, whereas cranes fly with their necks outstretched.

FAST WING FLAPPING
Fast, direct flight in a straight line with rapid wing beats is typical of pigeons, ducks, and seabirds with short wings, such as auks.

wing beats

INTERMITTENT WING FLAPPING
Woodpeckers and finches show an undulating flight of flapping interspersed with pauses in which they seem to bound up and glide down.

SLOW WING FLAPPING
Herons, gulls, and barn owls fly with a slow, steady, almost lazy-looking wing beat, allowing them to scan for food as they fly.

RANDOM WING FLAPPING
Aerial insect-eaters such as swifts and swallows usually have a random flight pattern, dipping and diving after insects as they hunt.

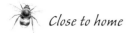

Feed the birds

Birds eat a range of foods, according to their species and the time of year. Some feed mostly on insects, especially in spring and summer when rearing their young, while others are mainly seed-eaters. Many birds gorge on berries and fallen orchard fruit in autumn and winter. You can enhance birds' survival by supplementing their natural diet with additional food.

BIG GARDEN BIRDWATCH

Why not get involved with the RSPB's Big Garden Birdwatch? This survey takes place annually during the last weekend in January. To take part you simply count the birds in your garden for an hour, note the highest number of individuals of each species you see at any one time, and log the results with the RSPB. For more information visit www.rspb.org.uk/birdwatch.

FEEDER PLACEMENT
Offer a variety of feeders in different locations around the garden. Place feeders within 2-3m of shrubs so that the birds can dash away if they see a predator such as a cat or hawk approaching.

caged feeder allows birds to feed on the ground without predators or scavengers getting in

windows kill birds that fly into them, locate feeders either within 1m or more than 10m away

elevated bird table fairly close to the house gives a good view of feeding birds

bird tables can be suspended from trees by chains

low bird tables may be used by timid species but ensure they are close to cover

tube feeders can be hung from tree branches or from brackets on walls

feeders on poles can be moved around the garden to find the birds' favourite feeding places

BIRD TABLES
Blackbirds are common visitors to bird tables. Offer a mix of flaked maize, black sunflower seeds, and peanut granules on your table.

USING CAMERAS

The key to successful bird photography is to be as unobtrusive as possible. Use a telephoto lens from the house, or turn the garden shed into a hide. Alternatively, place remotely operated cameras beside feeders or outside nest boxes to observe avian comings and goings. Some cameras are small enough to fit inside nest boxes so you can observe the growing family within.

NEST-BOX CAMERA
A small video camera mounted in a removable drawer in the roof of this box gives an excellent view of the nest inside.

MOTION SENSITIVE
Birds moving in front of the infrared sensor on this camera cause it to whirr into action, taking digital photos, videos, or a combination of the two.

Bird-mix recipe

To make fat balls, melt one-third suet or lard and mix well with two-thirds seeds, dried fruits, nuts, and oatmeal. Allow the mixture to set in an empty yoghurt carton or half a coconut shell, then hang it from your bird table.

Different bird feeders

Variety, both of feeder design and food types, is essential to attract the most birds. Some birds like to hang from feeders; some perch; others feed on the ground. Suitable foods include black sunflower seeds, niger seeds, flaked maize, peanuts, mealworms, and dried insects.

1 Small birds such as these goldfinches enjoy feeders with many perches specialized for holding small seeds. Niger seeds are small and black with a high oil content: a favourite of finches and siskins.

2 Nuthatches usually perch on tree trunks head-downwards, so a wire-mesh feeder is ideal for them.

3 Birds, such as this siskin, particulary need our help in cold weather. Ensure your feeders are replenished regularly and keep them clear of snow.

4 To attract great spotted woodpeckers, pack a coconut shell with suet and sunflower seeds and hang it from a branch or bird table.

Do's and don'ts

1. Use good-quality bird food from specialist suppliers.

2. Clear away stale or mouldy food.

3. Don't give leftover cooking fat, margarines, vegetable oils, or milk.

4. Never put out salted food or add salt to a birdbath.

5. Keep stored food dry.

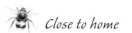

LADYBIRD

Wildlife garden

However big your garden, there are many things you can do to encourage wildlife. You will make a difference, and enjoy it, too.

Lavender in flower attracts butterflies and bees.

Why have a wildlife garden?

Many suitable habitats for wildlife have been lost or degraded over the years, through changing land use and development, but with a little effort you can provide your own safe places for wildlife to breed, shelter, and find food. There are plenty of activities that don't cost much, but give wildlife a helping hand whether you have your own large garden or a small balcony or window ledge.

Making a window box

If you don't have much outside space, create a mini nature reserve in a window box. Give interest and a food supply year round by choosing plants that bloom at different times. For example, spring bulbs attract early flying bumblebees, while summer flowers provide nectar for sun-loving butterflies. Many herbs, such as lavender and catmint, are favourites with honeybees and they also smell pleasant. Evergreens, such as rosemary and ivy, provide shelter for insects such as ladybirds throughout the winter.

1 Start by filling your window box with compost, then plant a variety of plants far enough apart to give them room to grow.

3 A very shallow saucer of water will attract water bugs and wildlife that may use it to drink or wash in. Water your window box regularly and give it some organic plant food, but avoid pesticides.

2 Cover the surface with gravel or bark to help the compost retain moisture during the summer, and to insulate the window box during the winter.

Bucket of life

A bucket of water left outside the back door will soon be teeming with larval insect life. Mosquitoes lay their eggs in still water. Once they hatch, the larvae mostly stay at the surface to breathe, but tap the bucket and they will wiggle underwater. You may also see small red organisms called bloodworms, which are the larvae of non-biting midges.

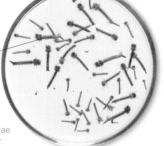

head end of the larva

TEEMING WITH LIFE
Mosquito larvae have mouth brushes on their heads for feeding on algae and bacteria in the water.

Bird bath

Birds require water to drink and wash in all year round. Place a bird bath close to bushes, but where the birds have good visibility. Clean and refill the bath regularly with fresh water, and break any ice. Water between 2.5cm and 10cm deep will suit a variety of species.

BATH IN THE AIR
Commercially produced bird baths include those that can be suspended from a tree.

SPLASHING AROUND
A pedestal bird bath, raised at least 1m above the ground, provides protection from hunting cats.

Helping the bees

Bees are vital for the pollination of many plants and food crops. They are in decline and need our help. Solitary bees nest in holes in the ground or in hollow plant stems. They seal an egg in a cell within the nest with some food for the larva to eat once it hatches. Give somewhere for the young to grow with a bought or home-made bee house. Place it in a sunny position and ensure rain can't get in.

BEE HOTEL

Make your own bee house from a bamboo cane, modelling clay, and a clay plant pot. Cut the cane into 20 equal lengths, according to the size of your pot. Bind the canes with tape and press the end of the bundle into the modelling clay. Wedge it into the pot with the open ends facing out.

Log pile

Replicate the valuable habitat of fallen trees with a pile of logs and branches in a shady part of your garden. This creates a damp, dark refuge and source of food for numerous animals, including beetles, centipedes, toads, frogs, and newts. Leave the wood to decay and you may see some intriguing fungi as it rots, especially if the logs are from different tree species. Add new logs periodically.

ROTTEN MEAL

Stag beetle larvae feed on rotting wood for up to five years before pupating over winter, emerging as an adult in spring.

Leaf piles and nesting boxes

Provide nesting materials for small mammals by raking dead leaves into a pile and putting out pieces of animal wool and hair, which they can use for lining their nest. You can also build or buy nesting boxes suited to hedgehogs or dormice for example. It is best to let a leaf pile rot into nutrient-rich leaf mould that you can later spread on the garden.

HELP A HEDGEHOG

A hedgehog might hibernate in a nesting box in a quite corner of your garden. Hedgehogs eat slugs, so don't put down slug pellets as these can harm them.

Look out for invertebrates, such as snails, in leaf piles.

COSY NEST

In addition to small mammals, hibernating toads may also use your leaf pile.

Bird and bat boxes

There are numerous different types of bird and bat box designed for nesting and roosting. You should consider the species you are hoping to attract when choosing which one to buy or build. Attach the box to a tree, post, or wall in a sheltered, quiet part of your garden. Clean it out in the late winter or early spring to remove abandoned nests and make room for new ones.

1 Position several bat boxes around the tree so they offer different temperatures during the day. Bats enter through a narrow slit in the base.

2 A more natural looking bird or bat box can be fashioned out of a section of a hollow tree branch, slotted into a hollow tree trunk.

3 Small birds, such as blue tits, require a box with a small entrance hole. Site the box well out of the reach of predators such as cats.

Compost dwellers

ANT
Ants often nest inside compost heaps. They also feed on some composting materials.

CENTIPEDE
Centipedes patrol the top layers of compost heaps, feeding on other insects and spiders.

EARWIG
Earwigs eat plant matter in compost. They use the pincers at their back ends to deter predators.

MAGPIE
Magpies and small mammals might scavenge food from your compost heap.

Composting
Compost is a mixture of decaying organic substances. Making a compost heap is a fantastic way of recycling your garden refuse and some of your household waste. You will also reduce the amount of rubbish going into landfill and produce an organic soil enhancer for your garden. The warm, moist environment attracts various animals to live and feed there. Add some old compost or soil to a new heap to introduce beneficial soil microbes and earthworms. Turn the pile regularly to add oxygen to it, but take care not to injure any animals.

RAW MATERIALS
Add fresh leaves, clippings, or food waste to the top of the pile. This provides a home for numerous insects, and will attract creatures that prey on them. You may even see birds picking over the top of your heap.

MATURING COMPOST
Bacteria cause most of the decay in a good heap. The heat they produce through digestion warms the heap and speeds up the composting process. Fungi, earthworms, and other invertebrates help, too.

FINISHED COMPOST
The compost should be ready after four to eight months, depending on the heap's temperature. Add finished compost to flower beds and vegetable patches to improve the structure of the soil and help your plants to grow.

Worms can burrow as deeply as 1.6m, but they surface after rain.

The different layers allow you to see how worms tunnel and process soil.

RECYCLING NUTRIENTS
Add your kitchen and garden waste to your home composting bin, and it will turn into valuable nutrients to put back into your garden.

Ingredients of good compost
To make good compost, you will need a mix of nitrogen-rich "green" ingredients such as fruit and vegetable peelings and grass clippings, and dry, carbon-rich "brown" matter such as leaves, straw, and wood shavings. Keep the heap moist and well aerated, and be careful not to add meat or fat to the compost to avoid rodent use.

WORMERY
Earthworms add oxygen and nutrients to the soil. You can build a wormery in a clear container to watch them at work. Make drainage holes in the bottom, then add alternating layers of sand, soil, compost, and leaf litter. Finally, put in some earthworms. Keep the wormery in a dark place and watch the tunnelling begin.

Making a pond in your garden

A pond is probably the most valuable asset to any wildlife garden. You should attract aquatic and semi-aquatic insects, including pond skaters, water beetles, water snails, dragonflies and damselflies, and amphibians such as frogs and newts. It will also provide somewhere for birds and mammals to stop for a drink. Stock your pond with native plants; be careful not to introduce non-native species that could cause damage. Avoid goldfish, too, as they eat water snails and tadpoles. Position your pond in a semi-shaded, sheltered location away from trees, and wait for the wildlife to arrive.

1 Dig a shallow hole with a sloping edge and shelved sections for access and gradients. Line the hole with a pond liner and temporarily weigh it down.

2 Cover the liner with sand, soil, and gravel. Ideally, fill the pond with rainwater from water butts, or allow it to fill naturally with rain.

POND IN A BARREL
Even if you have no garden, you can turn a container into a small pond. Steep sides cause problems for small creatures, however, so be sure to make access points.

3 Add plants, using bricks or stones if necessary to raise them to the required level in the water. A variety of submerged, floating, and emergent plants is recommended in order to attract as much wildlife as possible to your pond.

Water plants

OXYGENATING
Hornwort helps to keep the pond water clear, and also adds oxygen during photosynthesis.

DEEP WATER
Water hawthorn leaves float on the surface of ponds, but its bulbs can be planted up to 60cm deep.

MARGINALS
Water forget-me-nots grow at the pond margin, giving resting places for insects and protection for wildlife.

FLOATING
Frog's-bit plants float without putting down roots. They provide valuable cover on the pond's surface.

SUNKEN CONTAINER
Watertight containers, or even old baths, can make ponds if you sink them into the ground. Clear out any leaves that fall in.

Butterflies and moths

You can make your garden more attractive to butterflies and moths, which can also be found in many other habitats worldwide.

Butterfly or moth?

Butterflies and moths both belong to the insect order Lepidoptera, meaning "scale wings", which has about 180,000 species. Most butterflies are diurnal, while many, though not all, moths are nocturnal. Many moths appear drab in comparison to more brightly-coloured butterflies – their colours camouflage them while they are resting. Moths usually rest with their wings outstretched, while butterflies fold their wings vertically over their backs, unless they are basking in the Sun.

long, sensitive antenna

brightly coloured wing

broad abdomen

SMALL ELEPHANT HAWKMOTH (SPHINGID)
These moths feed on the nectar of flowers such as honeysuckle and rhododendron. Their caterpillars eat bedstraws.

forewing

hind wing

SWALLOWTAIL BUTTERFLY
There are over 600 species of the swallowtail butterfly family (Papilionidae) worldwide. Particularly prevalent in tropical regions, their name derives from the pointed tips of the hind wings, which resemble the forked tail of a swallow.

tail on hind wing

METAMORPHOSIS

Butterflies and moths undergo a spectacular transformation, or metamorphosis, from leaf-munching caterpillars to winged adults. During a process called pupation, the juvenile cells in the caterpillar's body are repurposed, while the undifferentiated cells, known as imaginal discs, divide, elongate, and become specialized. Pupae are encased in a hardened shell called a chrysalis. Some moth caterpillars spin a protective silk cocoon around this before pupating.

1 Butterflies and moths usually lay their eggs on plants. Look for clusters of tiny, hard-shelled eggs that are "glued" to the upper or lower surface of a leaf.

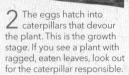

2 The eggs hatch into caterpillars that devour the plant. This is the growth stage. If you see a plant with ragged, eaten leaves, look out for the caterpillar responsible.

pupa has a protective, hard case

3 When the caterpillar is fully grown, it looks for a safe place to pupate. Its skin splits to reveal the pupa, inside which it transforms into an adult.

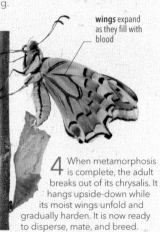

wings expand as they fill with blood

4 When metamorphosis is complete, the adult breaks out of its chrysalis. It hangs upside-down while its moist wings unfold and gradually harden. It is now ready to disperse, mate, and breed.

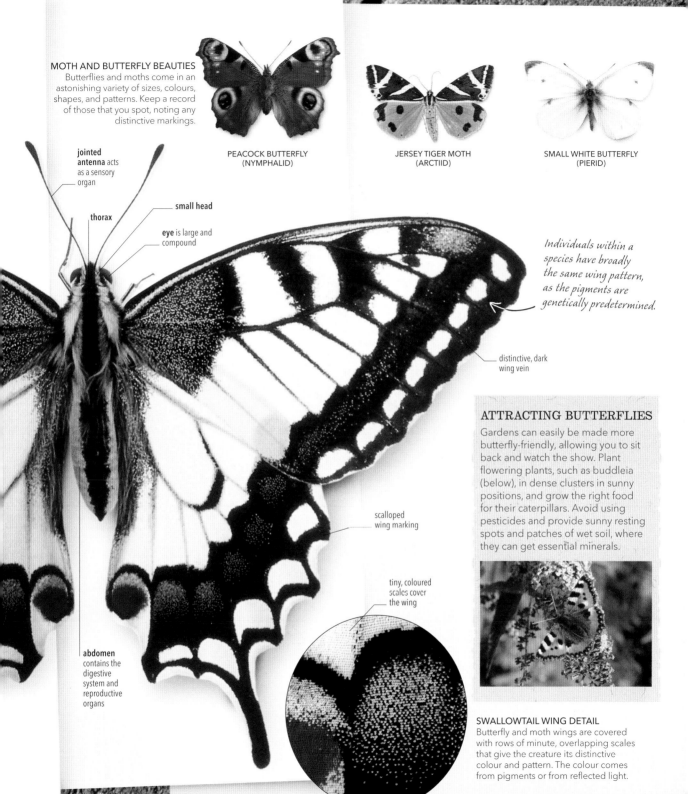

MOTH AND BUTTERFLY BEAUTIES
Butterflies and moths come in an astonishing variety of sizes, colours, shapes, and patterns. Keep a record of those that you spot, noting any distinctive markings.

PEACOCK BUTTERFLY
(NYMPHALID)

JERSEY TIGER MOTH
(ARCTIID)

SMALL WHITE BUTTERFLY
(PIERID)

jointed antenna acts as a sensory organ

thorax

small head

eye is large and compound

Individuals within a species have broadly the same wing pattern, as the pigments are genetically predetermined.

distinctive, dark wing vein

scalloped wing marking

tiny, coloured scales cover the wing

abdomen contains the digestive system and reproductive organs

ATTRACTING BUTTERFLIES
Gardens can easily be made more butterfly-friendly, allowing you to sit back and watch the show. Plant flowering plants, such as buddleia (below), in dense clusters in sunny positions, and grow the right food for their caterpillars. Avoid using pesticides and provide sunny resting spots and patches of wet soil, where they can get essential minerals.

SWALLOWTAIL WING DETAIL
Butterfly and moth wings are covered with rows of minute, overlapping scales that give the creature its distinctive colour and pattern. The colour comes from pigments or from reflected light.

Farm and field

Farmland may be a modified aspect of our landscape, but it is often the most accessible type of countryside for people to explore. And while it's true that much modern farming and forestry is destructive to native species, in some places farmers have managed their land less intensively and have actually created new habitats that are richer than some natural environments – sometimes allowing small groups of species to prosper artificially. In the UK, for example, both poppy fields and skylarks have flourished on farmland, and who would not be excited by a walk across a flower-filled "unimproved" meadow? Indeed, these are all national – and natural – treasures.

ORCHID

Pasture

If managed sensitively, with techniques such as rotational grazing, livestock can help maintain grassland habitats for wildlife. Without livestock to crop vegetation, strong grasses and shrubs would take over, and wild flowers such as orchids would be lost. Grasslands are vital to butterflies and birds such as the stone curlew, woodlark, and nightjar. Flooded water meadows provide a seasonal refuge for water birds.

SNOWY WAX-CAP MUSHROOMS

EUROPEAN MOLE

Farm and field

Wildlife can struggle to find a niche in today's agricultural landscapes. When farmed in a wildlife-friendly way, however, they are a mosaic of habitats: pasture, crop fields, and meadows interspersed with hedges, ditches, and woodland.

Arable land

Arable fields, where crops are grown, are often bleak monoculture, where few animals live. However, organic or other sensitive farming techniques provide wildlife with food and habitats. Measures to help wildlife include leaving crop remnants in the fields during the winter after harvesting, rotating the crops, reinstating lost hedgerows, and leaving wide field margins.

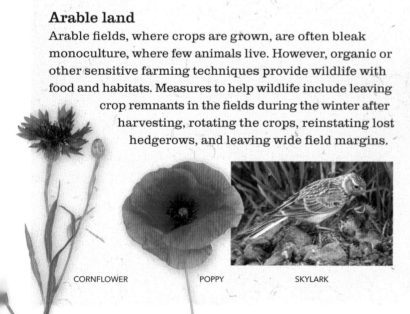
CORNFLOWER   POPPY   SKYLARK

Field boundary

Much of the value of farmland to wildlife is found at field edges, where machines and chemicals are kept at bay. Hedgerows, ditches, and stone walls are all types of field boundary. They can provide wildlife, such as songbirds and reptiles, with food, shelter, and corridors through which they can safely travel.

COMMON SHREW

SONG THRUSH

Hay meadow

Wild flower-rich hay meadows aren't just beautiful – they support a myriad of insects that feed on nectar, including bumblebees, honeybees, and butterflies. Sadly, traditional hay meadows are now few and far between because of changes to cutting regimes and the use of herbicides. This has resulted in a reduction in insect numbers.

BUTTERCUP

BARN OWL

Barns and outbuildings

Farm buildings provide homes for barn owls, swallows, bats, and rodents such as mice. When a little rundown and seldom disturbed, barns and outbuildings are excellent places to watch wildlife. An open window or gap in a door allows animals to enter and find shelter.

Old farm buildings can be sympathetically restored to ensure they continue to welcome wildlife in the future.

HARVEST MOUSE

69

Nature's highways

Despite being cultivated, farmland can play host to an abundance of wildlife. Field boundaries provide animals with safe passage.

Field boundaries

Different farming systems have used different ways of defining fields according to their requirements, available materials, and the climate. Traditional methods include hedges, ditches, and stone walls, all of which provide opportunities for wildlife to inhabit and travel through the agricultural landscape. The loss of field boundaries from the environment, whether by removing hedgerows to create larger fields, or replacing stone walls with fences, is detrimental to the wildlife that depends on them.

VIEW FROM ABOVE
Look down from an airplane and you can see how fields are separated by hedges, tree lines, ditches, or walls. A mixture of boundary types creates the most wildlife-rich farmland.

Hedges

Hedges are rows of shrubby plants grown to demarcate fields, the edges of lanes, or settlements. You can estimate a hedge's age by counting its plant species – some hedges can be hundreds of years old. Many different types of animals and plants thrive in the shelter hedgerows create, including birds, insects, mammals, wildflowers, mosses, and fungi.

ADDER
Various reptiles find food and shelter in hedges, including snakes such as the adder (common viper).Adders are venomous, so don't get too close.

Ditches

Ditches drain low-lying land for agriculture, but they may also be home to insects, such as water beetles and dragonflies; water birds; newts and other amphibians; and mammals, including otters and water voles. Ditches must be managed to prevent silt blockage or clogging with vegetation. They may also be adversely affected by fertilizer and pesticide run-off from adjacent agricultural land.

AMERICAN MINK
The American mink is native to North America but has been introduced to Europe. You may spot one in an agricultural ditch. Mink hunt and kill native wildlife, including water voles.

FIELD VOLE
Field voles are tiny rodents native to Northern Europe that require grassy habitats. They may hide in hedges to escape being noticed by passing birds of prey – or run to them for protection.

MOORHEN
The moorhen is one of the most likely examples of the waterbirds you might find using farmland ditches.

GERMINATION IN ACTION

Germination is the process by which a plant seed begins to grow into a plant. It can be monitored at home by placing some seeds on moist paper towel in a clear sealed plastic bag, and then leaving it for a week or so at room temperature. Seeds are contained within a fruit – or, in the case of beans, a pod, which splits open. Each bean is a seed that will become a new plant, provided that conditions such as temperature, water availability, and light intensity are suitable. In cactus hedges, these can be ideal for black-bean germination.

1 The black bean seed contains the plant embryo and a store of starch and protein surrounded by a seed coat.

hilum scar where seed was attached to parent plant

radicle simple first root of plant

2 Germination begins, and an embryonic root called the radicle emerges from the seed and grows downwards.

cotyledons first plant leaves, which begin process of photosynthesis

3 The seed leaves, or cotyledons, are pushed towards the surface by growth in the embryonic shoot, or hypocotyl.

secondary roots begin to anchor plant, and take in water and nutrients

Disused railway lines

Reclaimed by nature after the trains stopped running, old railway lines form a network of linear nature reserves. Often rich in flowering plants and the invertebrates that pollinate them, these peaceful corridors are often excellent places to explore for butterflies and other insects. Small mammals use these natural highways to move from one area to another, and bats often follow them as they fly from their roosts to forage after dusk.

KESTREL
The kestrel is a falcon that hunts small mammals living in low vegetation. It locates them by sight as it hovers.

FLOWER POLLINATOR
Butterflies such as the brown argus are important pollinators. As they move from flower to flower in search of nectar to drink, they also transfer pollen.

Stone walls

Stone walls have long been used to enclose fields or to terrace sloping agricultural land – for example, in Mediterranean olive groves. If built without mortar or cement, they provide plenty of nooks and crannies for plants and animals. Look for insects, reptiles, and amphibians that make their homes in stone walls, and study the surface of the stone closely to see lichens growing.

SUN LOUNGER
Reptiles, such as this wall lizard, bask on and hide between the warm rocks of stone walls.

MOUNTAIN MADWORT
This plant thrives in dry, rocky soil in Europe and Asia. Wild flowers like this may seed in wall crevices.

Shape shifters

The sight of starlings flocking is a thrilling spectacle. Moving as one, the birds morph gracefully over open land.

Watching flocking birds is one of nature's great displays. If you are fortunate you might see a flock of European starlings – gregarious birds that feed and roost together in the hundreds or thousands. Flocks are at their largest in autumn and winter. Look for them over farmland, especially grazing marshes, in the afternoon and early evening. After a busy day feeding, the birds fly together for up to half an hour on their way to nearby roosts in woodland, reedbeds, or urban areas. Flocking, also called murmuration, helps the birds to avoid predation by making it difficult for birds of prey to pick out individuals. Groups of European starlings also exhibit an interesting behaviour when feeding. As the lead birds move forward over the ground searching for insects, birds from the rear fly up and over to land in front. This behaviour is called "roller feeding". During spring and summer, the starlings feed in open farmland, probing insects and larvae from the ground. In this way they aid farmers by controlling a number of crop pests. In the autumn and winter they turn to grain and fruit.

FLOCKING TECHNIQUES

It was previously thought that European starling flocks maintained their cohesion because each bird kept a set distance from its nearest neighbour. However, new research has shown that each starling tracks the location of six or seven other birds within the flock, allowing it to make the necessary adjustments to its flight to keep the flock together. In this way the flock can expand and contract while maintaining cohesion and avoiding the risk of a lone bird breaking away.

AVOIDING A PREDATOR
The flock weaves about, splits, and reforms in the air in an effort to outwit predators such as peregrine falcons.

Beetles

Beetles can be found in almost all habitats, from deserts and ponds to the tops of mountains. They are thought to represent one third of all insects.

Beetle diversity and distribution

Beetles (Coleoptera) are arthropods, a major group of invertebrates that also includes arachnids and crustaceans. They have jointed legs and front wing cases (elytra) that cover and protect the more delicate hind wings. There are believed to be well over 350,000 species of beetle worldwide, with many more species yet to be discovered. Beetles play an important role in the natural world – they recycle nutrients by helping to break down animal and plant waste. Most species are herbivorous.

POTATO-EATER
Native to North America, the Colorado beetle (family Chrysomelidae) has been introduced widely. It can cause damage to potato crops.

Beetle varieties

There are currently around 188 different families of beetles, but their classification is constantly being reviewed. They range in size from tiny species smaller than a millimetre to giants, such as the titan beetle of the Amazon rainforest, that are nearly 20cm long. Here are some common families.

LADYBIRD (COCCINELLIDAE)
Ladybirds are small, domed beetles. The most familiar pattern is red with black spots, but other colourways also exist.

ROVE BEETLE (STAPHYLINIDAE)
The carnivorous rove beetles have long, flexible abdomens visible beneath short wing cases (elytra).

WEEVIL (CURCULIONIDAE)
Weevils, or snout beetles, are small plant-eaters with bent, clubbed antennae. Over 60,000 species are known.

SCARAB BEETLE (SCARABAEIDAE)
Scarab beetles vary enormously in colour, shape, and size. All have a distinctive club at the ends of their antennae.

LONGHORN BEETLE (CERAMBYCIDAE)
These beetles are so-called because they possess antennae that may be as long as, or longer than, their bodies.

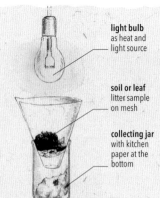

- **light bulb** as heat and light source
- **soil or leaf litter sample** on mesh
- **collecting jar** with kitchen paper at the bottom

MAKING A BERLESE FUNNEL

A Berlese funnel lets you see the variety of insects that hide in soil and leaf litter. Put some soil on a piece of mesh in a funnel, paper cone, or half an empty plastic bottle and suspend it over a jar with some kitchen paper in the bottom. Shine a lamp onto the sample from at least 10cm above. The heat and light coax any insects to move through the mesh into the jar. Ensure you let them go.

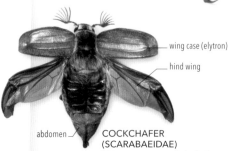

wing case (elytron)

hind wing

abdomen

COCKCHAFER (SCARABAEIDAE)
This common cockchafer's wings are outstretched for flying. Their antennae fan out to sense air currents.

mouthpart
(mandible)

antenna

head

leg
joint

foreleg

thorax

wing case
(elytron)

GROUND BEETLE (CARABIDAE)
There are almost 30,000 species
of the carnivorous ground beetle.
They are usually black, metallic blue
or bronze, or iridescent green. They
use a comb-like structure on their
forelegs to clean their antennae.

BURROWING DOR BEETLES
Like many scarabs, dor beetles feed on
animal dung. They excavate burrows directly
beneath piles of dung, especially that of
cattle and horses. The burrows are generally
15-200cm deep, sometimes with side
chambers, and the beetles fill them with dung.
Unlike some scarabs, they carry the dung
underground rather than rolling balls of it. The
females lay their eggs in, or close to, the dung
in the burrow. When they hatch, the larvae have
a ready supply of food.

AGING ROCKERS
If a dor beetle rolls onto its back, it can usually
right itself by rocking from side to side until it
builds the momentum to turn over. When they
get old, however, these beetles lose this ability
and cannot get themselves upright again.

*Dor beetles have
toothed front legs
that help them dig.*

Beetles in combat
Stag beetles (family Lucanidae) are so named because
their large mandibles or mouthparts resemble the
antlers of stags. Male stag beetles also fight for access to
females. Fights are likely to occur close to places where
females might lay eggs, such as decaying tree stumps.

1 When two males
are interested in
the same female, they
approach each other
threateningly in an
attempt to gauge their
rival's size and strength.

2 If threats alone
are insufficient, the
beetles grapple with
their mouthparts as each
attempts to lift the other
off its feet by grasping it
around the middle. The
winning beetle then drops
the loser to the ground.

3 Although a beetle
might end up on its
back, these fights seldom
result in death or injury.
They are more a show
of strength to establish
a hierarchy for mating.

Open farmland

Farming need not replace nature. If farmland is sensitively managed, wild flowers and animals can thrive at its edges, or within the fields themselves.

Exploring farmland

Farmland can be a great place for wildlife-watching, but take care before exploring it. If there are no public rights of way (see p.40), you must get permission from the landowner before going on private land. Always keep to footpaths, bridleways, or field edges to avoid damaging crops. Don't startle livestock, always leave gates as you find them, and avoid any fields that have been recently sprayed with herbicide or fertilizer.

HARVEST TIME
The mechanization of farming during the last century has meant more arable land – but loss of wildlife habitat.

MOLE HILLS
Little heaps of earth in pastures are a sure sign that moles are tunnelling beneath, looking for earthworms.

Wildlife

Environmentally friendly farming can benefit insects such as bees and butterflies, and birds and mammals, including deer, rabbits, foxes, and bats. A good way to observe farmland wildlife is to choose a position downwind on a footpath where you can watch from behind a tree or hedge without giving away your presence or disturbing any animals. Use binoculars to help you spot and identify species. Wildflowers, such as poppies, cornflowers, and field pansies, can be found in arable fields, but only at the edges if herbicides have been used on the crops.

SHORT-TAILED VOLE
Voles are tiny rodents that feed on grasses. Short-tailed voles live in farmland, meadows, and forest clearings in much of Europe.

ACTIVE ADAPTATION
Try looking for banded snails in fields and hedges. This species comes in more than one coloured form. You may notice more dark-shelled snails on brown leaves and yellowish snails in grass. Each is camouflaged from predators, such as thrushes.

BROWN-LIPPED

WHITE-LIPPED

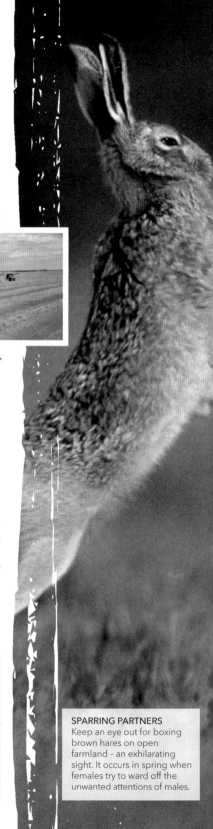

SPARRING PARTNERS
Keep an eye out for boxing brown hares on open farmland – an exhilarating sight. It occurs in spring when females try to ward off the unwanted attentions of males.

Farmland birds

Many birds that inhabit farmland nest on the ground, including the skylark, lapwing, corn bunting, and grey partridge. They are difficult to see on their nests, but you may catch sight of them in flight or when feeding. Also look for birds of prey, such as kestrels or barn owls, hunting for small rodents. In Europe, farmland birds are declining in numbers, as are grassland birds in North America. This is thought to be largely due to changing agricultural practices.

LAPWING
Birds such as lapwings breed in arable fields and meadows. They eat insects and worms.

However, a number of major conservation efforts are under way to help reverse the decline, such as providing uncultivated areas within fields where birds can nest and forage, and growing flower-rich margins at field edges to attract insects.

GREY PARTRIDGE
Adult partridges feed on grain in cereal fields. Their young eat insects in the field margins.

COMMON PHEASANT
Pheasants originally from Asia are reared and released in their millions for people to shoot. Such numbers greatly impact native wildlife.

Cowpats

Untreated cattle dung supports over 200 species, mainly because cowpats are a great source of food for worms, flies, beetles, and springtails. In return, these tiny creatures do a great job of removing cowpats and recycling their nutrients back into the soil. Dung beetles also help keep numbers of harmful flies, such as horn flies and face flies, low in farmland by outcompeting them in cowpats. Sadly, drugs and other chemicals fed to cows can destroy these communities.

SPRINGTAIL

DUNG FLY

DUNG BEETLE

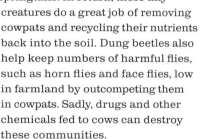

COWPAT SOCIETY
On farms that use fewer drugs and pesticides, you can see male common yellow dung flies establish territories on fresh dung and wait for females. After mating, the females lay eggs on the dung's surface.

A healthy dung insect community will remove a cowpat from the soil's surface within 24 hours.

Farmland close-up

Farmland varies hugely around the world and the wildlife inhabiting it depends upon the type and method of agriculture being practised. Wherever you are, look along field margins for signs of life.

SWEET CHESTNUT

NUTS

Autumn brings a rich harvest of fruits, nuts, and grains.

THISTLE

TRAVELLER'S JOY

WOOD BETONY

Look for blossoms adding colour to fields in spring and seedheads in late summer.

HEDGE PARSLEY

MAIZE

WHEAT EARS

CRAB APPLE BLOSSOM

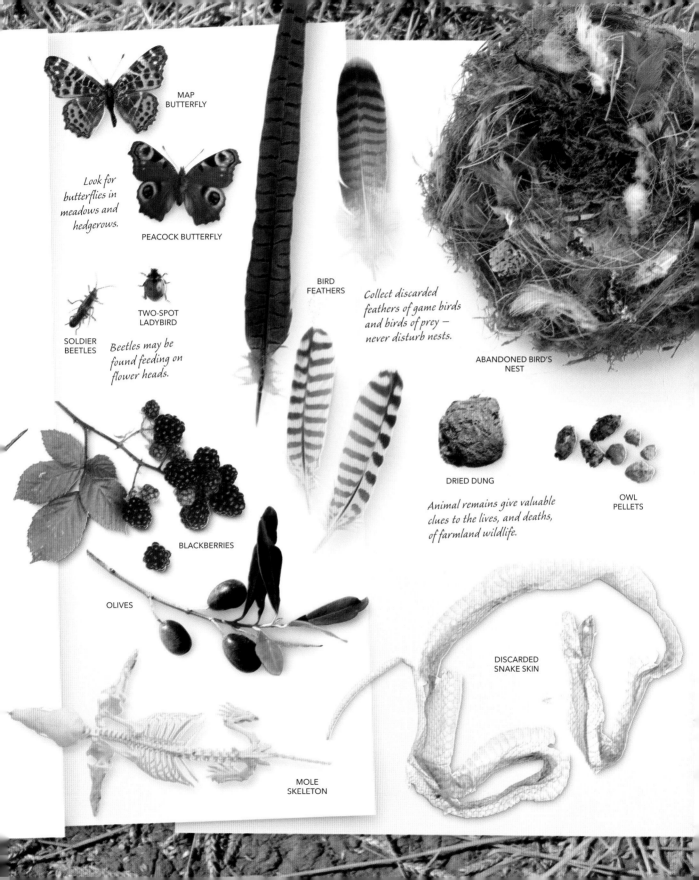

MAP BUTTERFLY

Look for butterflies in meadows and hedgerows.

PEACOCK BUTTERFLY

TWO-SPOT LADYBIRD

SOLDIER BEETLES

Beetles may be found feeding on flower heads.

BIRD FEATHERS

Collect discarded feathers of game birds and birds of prey — never disturb nests.

ABANDONED BIRD'S NEST

BLACKBERRIES

DRIED DUNG

Animal remains give valuable clues to the lives, and deaths, of farmland wildlife.

OWL PELLETS

OLIVES

DISCARDED SNAKE SKIN

MOLE SKELETON

Forest

You walk over grassland or heath, but you walk in a forest. It envelops you, and this points to a significant difference – there are many more opportunities here for life to exploit distinct niches. Thus, forests support our richest biodiversities. One oak tree, for example, may provide the life-support systems for more than 1,000 species of invertebrate. Forests are complex places, where the connections between the species that form communities are still revealing surprises. Yet forests, from the vast northern taiga to modest copses in the English countryside, can also be difficult habitats to explore –you can't see the life for the trees. You will need a lifetime of patience to uncover their secrets.

Deciduous woodland **p.82**

Coniferous forest **p.106**

Ancient woodland

A few forests have remained almost unchanged by human influence. In an area that has been consistently wooded for hundreds of years you may find rare plants, such as orchids growing on chalk in beech groves. Birds, such as owls, and mammals adapt well to modern woods, but older forests often attract a wider array of animals than younger woodlands, including more specialized insects and plants.

TAWNY OWL

OAK BEAUTY MOTH

Deciduous woodlands

From the tropics to temperate zones, broadleaved woods are beautiful habitats. Deciduous forests that are leafless in winter are found mainly in moist, relatively cool habitats, and at higher elevations closer to the tropics.

WILD GARLIC

Beech forest

Beech forests glow a brilliant green in spring and burn bright with glorious autumn colour. The dense canopy keeps the forest floor in deep shade so few shrubs grow here, but you can find plants such as wild garlic and abundant fungi springing up on the forest floor. Birds sing in the trees, insects flourish in the leaf litter, and you can find squirrels, foxes, and – if you're quiet – badgers among mammalian residents.

BEECH LEAF

82

Parkland

Parkland is largely artificial in Europe, often associated with estates around manor houses. The shrub layer is often weak or absent, which means that you won't see as many butterflies and forest flowers as on a heath or grassland area, however old, rotting trees still attract birds such as little owls, nuthatches, and woodpeckers, as well as insects, and you may encounter mammals such as deer or squirrels.

FOXGLOVE

HORSE FLY

Temperate rainforest

Temperate rainforests, which grow along the coasts of northern Spain, Ireland, Scotland, and Norway, are even more threatened and fragmented than most tropical rainforests. Yet they have a great wilderness value – mature forest is often ancient, with a covering of spring flowers or ferns, while open glades encourage a dense growth of thick, tangled shrubs at their edges – all home to a host of wildlife.

MARSH FRITILLARY

TREE LUNGWORT LICHEN

BLACKBERRIES

OAK LEAVES
AND ACORNS

PURPLE
EMPEROR

Remember to look up for birds and squirrels in the trees.

A walk in the woods

A deciduous wood is packed with a myriad of sights, smells, and sounds. Tall trees envelop you, shielding you from the interference of the human-made world.

The character of the wood changes dramatically with each season, so visit throughout the year to see how the forest develops. In spring, the wood comes alive with colourful flowers, insects, and birds. Trees are in full leaf during summer, providing food and shelter for insects in the canopy and covering the forest floor in shade. Look for fungi in autumn and for leaves of deciduous trees, which turn beautiful shades of yellow, orange, and red. In winter, the leaf litter is still a place of abundant activity.

BEECH
NUT

Look at fallen branches, which create food and homes for insects – the gaps they leave behind let sunlight shine in.

SOLOMON'S
SEAL

Sift through rotting leaves to find a thriving habitat of tiny plants and animals.

HAWKWEED

84

BEECH
LEAF IN
AUTUMN

*Look for holes in
trees — many birds
nest in them.*

COMMON WOLF
SPIDER

*Summer leaves
hide birds and
squirrels — listen
for their calls.*

HART'S TONGUE FERN

*See how bracken thrives
where sunlight reaches
through to the ground.*

FLOWERING
MOSS

Living space

From the roots to the canopy, trees provide living space for a wide variety of wildlife, thus increasing the habitable area of a forest.

Forest builders

Just a moment spent looking at a tree reveals that they abound with life. They provide food and shelter for animals both large and small, from the microscopic organisms in rotting leaves to the birds in the highest branches. Each tree creates a host of niches at every level – the larger the forest, the more diversity. Because the forest is so interdependent, it functions in much the same ways as a single living entity, with each species being reliant on another species for its survival. At the bottom of this chain is the soil.

FOREST WEB

Plant growth in the forest produces enough food to support three levels of consumers, from small herbivores to carnivores and large omnivores.

tree →

beetle, worm, finch, mouse, vole, deer →

fox, jay, owl, hedgehog →

wolf

MEASURING UP

Trees are the Earth's largest living organisms. Measuring a tree and counting its annual growth rings can provide more than dimensions and age – it can also reveal how environmental conditions have affected it over the years. The growth rings on this tree are off-centre, suggesting that one side of the tree may have been exposed to harsh, windy conditions.

— rapid growth on sheltered side

CROSS-SECTION FROM FELLED TREE

GIRTH GROWTH

The girth (circumference) of a tree increases each year. Measure it with a tape measure at least 1.5m (5ft) above the ground. Keep a record of your measurements and compare your readings over a number of years to see how the tree is growing.

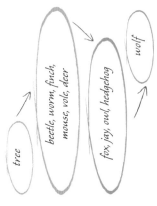

Tree levels

PURPLE HAIRSTREAK

5 CANOPY

Treetops create a canopy, letting through different levels of light according to the type of tree. Some animals, such as birds, butterflies, martens, and dormice, are canopy specialists, each choosing different "storeys" for different needs. Birds such as thrushes feed and sing in the canopy, but nest at lower levels. Butterfly caterpillars may feed in the canopy, while adults feed from flowers far below.

MISTLE THRUSH

GREY SQUIRREL

5

4 SHRUB
Shrubs and saplings grow beneath some trees, for example holly grows under an oak to create a shrub layer.

VIBURNUM

3 FIELD
Herbs and low-growing flowers, ferns, and mosses create a field layer close to the ground.

COLUMBINE

2 LEAF LITTER
Rotting leaves from past seasons create leaf litter, which has a thriving wildlife system of its own. Oak and sycamore leaf litter is especially rich.

1 SOIL
Humus from rotting foliage and dead wood mixes with underlying minerals to create the soil on which all life depends.

EARTHWORM

Tree shapes

Individual trees can develop unusual shapes, especially if they are growing in harsh conditions. Look for short, gnarled, and stunted trees at high altitudes, cold and windy areas, or poor soils. In dense forests you might find very tall and straight trees, while coastal trees tend to bend away from prevailing winds as twigs and branches on the exposed side die or fail to develop. If you come across trees with bark, leaves, and branches stripped to a specific level it may be damage from browsing animals.

SPECIES SHAPES

The outline of a mature tree can help you to identify it. With or without leaves, each species has its own characteristic shape based on the number and thickness of the smaller twigs and the angle at which they grow from the main branches.

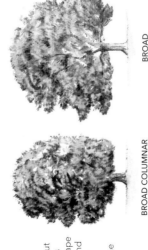

BROAD
(BEECH)

BROAD COLUMNAR
(SWEET CHESTNUT)

SLENDER AND GNARLED
(SESSILE OAK)

COLUMNAR
(HOLLY)

NARROW AND WEEPING
(SILVER BIRCH)

CONICAL
(ALDER)

SPREADING
(HORNBEAM)

COLUMNAR TO SPREADING
(WILD CHERRY)

87

Leaves

Leaves are the crowning glory of most plants, they convert energy from the Sun – the basis of all life on Earth.

How leaves work

Plants collect nutrients by photosynthesis. Chlorophyll in leaves uses sunlight to convert carbon dioxide from the air into starches and sugars. As leaf pores (stomata) take in carbon dioxide, water is lost by evaporation. Nutrient-rich sap rises, pulling water in from the soil via the roots in a process called transpiration.

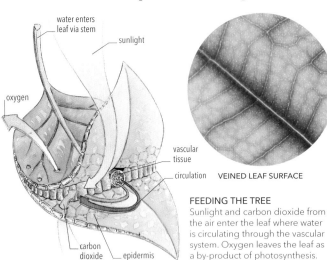

VEINED LEAF SURFACE

FEEDING THE TREE
Sunlight and carbon dioxide from the air enter the leaf where water is circulating through the vascular system. Oxygen leaves the leaf as a by-product of photosynthesis.

CHLOROPHYLL

Chlorophyll is an essential pigment contained in plants. It is responsible for the great variety of greens you can see in different kinds of plant and achieves this by mostly absorbing blue and red light and reflecting green light. Known as a "photoreceptor", chlorophyll is essential for photosynthesis. In winter, when days are shorter, photosynthesis is not as efficient and chlorophyll is withdrawn from the leaf cells, leaving them an orange-brown. Trees shut down and survive on stored starches until spring.

LEAF SURFACE
A waxy cuticle reduces water loss and tiny hairs control evaporation. When old leaves pit and mottle they lose this function.

young leaf
old leaf

Leaf structure

You will notice most conifers have needle-like leaves, but broadleaved deciduous tree leaves have a stiff midrib and a broader blade with supporting side ribs. A stalk provides mobility, reducing wind damage.

Leaf shape

Leaves come in a great variety of shapes and sizes. They may be simple or compound, with a number of leaflets such as "palmate" (hand-shaped) and "pinnate" (several leaflets in a row). Teeth, notches, points, and lobes encourage water to gather and drip away.

OVAL (ELLIPTIC)
TEARDROP (ARISTATE)
ROUND (ORBICULAR)
COMPOUND (PEDATE)
HEART-SHAPED (CORDATE)
ELONGATED (LINEAR)

Arrangement

Look for leaves with different arrangements. Some are solitary, while others are clustered. Where they have stalks they can be opposite each other or they can alternate.

branch from same point
OPPOSITE

branch alternately on each side
CLUSTERED
branch from central point
ALTERNATE

Colour varieties

Leaves come in a variety of colours. Some have a red pigment overlying the chlorophyll, creating "copper" varieties. Others have yellow pigments or appear almost white. Dying leaves in the autumn make sugar, which is trapped as a red pigment.

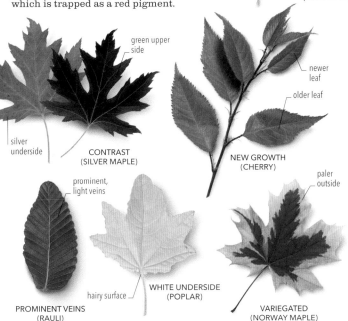

AUTUMN RED
(RED MAPLE)

green upper side

silver underside

CONTRAST
(SILVER MAPLE)

newer leaf

older leaf

NEW GROWTH
(CHERRY)

prominent, light veins

hairy surface

PROMINENT VEINS
(RAULI)

WHITE UNDERSIDE
(POPLAR)

paler outside

VARIEGATED
(NORWAY MAPLE)

Margin and texture

Smooth, waxy leaves tolerate dry or harsh conditions, while others are coarse with rough surfaces. The great variety of leaves are adapted to different habitats.

SMOOTH WAVES
(BEECH)

LOBED
(OAK)

COARSELY TOOTHED
(HORNBEAM)

SPINY
(HOLLY)

SMALL TEETH
(BIRCH)

CRINKLED
(ELM)

BLACK FOREST FALL

The Black Forest in Germany is one of the best places to visit to see a spectacular leaf fall. Most leaves live for about six months, before dropping in a process called abscission, and are replaced every year. A large, mature tree may have around 250,000 leaves. Cold nights reduce the transportation of sugars around the tree and they become trapped in the leaves. Along with the lack of chlorophyll, this creates the reds and yellows of a glorious fall.

Beech trees create some of the most vivid colours.

Seasonal change

A new leaf is usually a vivid, light green. Weeks of hard work, plus the effects of wind, rain, and temperature fluctuations, gradually hardens the leaf and makes it duller and darker with age, until it finally begins to "turn" and decompose.
Try photographing leaves from one plant throughout the year, to compare the colours as the year progresses.

LIFE CYCLE
Most leaves start life as bright green, but change colour as they age and begin to degrade.

as chlorophyll fades away, it reveals the colours of other pigments in the leaf

red anthocyanin pigment gives the leaf its distinctive red colour

Forest floor

The forest floor in spring can be stunning, with swathes of flowers, while in autumn, it is a scene of decay – yet still full of life.

Exploring leaf litter

Leaf litter is full to the brim with tiny animals such as insects, spiders, and salamanders. These are exploited by bigger ones, such as badgers, thrushes, and woodcock. A careful look at the leaf litter will reveal a wealth of wildlife. Identifying leaf litter flora and fauna is a specialist task, but you can get an idea of what is about in your local wood.

FINDING CLUES
Distinctive droppings or tooth marks on nibbled seeds offer clues to a woodland's inhabitants.

MOISTURE LOVER
Small, slender, and moist-skinned, salamanders, such as this fire salamander found in much of Europe, need a damp environment to survive.

SNAIL TRAIL
Snails move on a muscular foot, coated with mucus to reduce friction and protect against sharp surfaces, leaving a slimy trail.

INSECT HUNTER
All sorts of insectivores, including shrews such as this pygmy shrew, can be found in leaf litter rich with insects, worms, and spiders.

woodlice have tough, hard skins to retain moisture

earthworms process vast amounts of soil and leaf matter

beetles and their larvae feed in dead leaves

fallen leaves provide food, shelter, and a vital moist environment

Take a clear plastic box and a hand lens to examine small insects.

leaf litter spiders ambush prey, rather than relying on webs in this mobile environment

USING A POOTER

The pooter has a short tube with gauze at one end, and a long open tube. Put the long tube over an insect, then suck sharply through the short tube. The insect should shoot into the cylinder. The membrane stops you sucking it into your mouth so you're not going to swallow it, but always make sure you're sucking the right tube.

BUG COLLECTOR
Pooters are devices for catching insects, so you can identify them and release them unharmed.

Forest flowers

Flowering plants face many challenges in woodland – trees demand a huge amount of water from the soil, and, when in full leaf, cast a deep shade on the forest floor. Flowers have evolved many ways to deal with such problems. Some, such as primroses, wood sorrel, and wood anemones, flower very early in Europe, before the trees are in full leaf. Others, such as toothwort and several kinds of orchids, parasitize other plants. Many simply grow on the sunny edges of woodland or forest glades.

FLOWER CARPET
In Europe, flowers such as bluebells carpet forest floors. Like lesser celandine, they bloom in early spring.

slugs feed on forest floor debris and fungi, which helps to disperse seeds and spores

saddle-shaped mantle has a respiratory opening

A small garden trowel will help lift a soil sample as well as leaf litter.

earthworms leave curly "casts" of decomposed matter

larvae developing underground are sheltered from cold winters

LESSER CELANDINE

CREATING TOPSOIL
A cross-section of the leaf litter and upper soil layer on the forest floor reveals a host of animals and plants that help digest the fallen leaves.

Nature's recyclers

The world of rotting logs and leaves appears to be dead but is actually alive with miniscule animals, but not all of them are visible to the naked eye.

A wood is a celebration of life's abundance and even so-called "dead wood", such as this rotting log, is actually alive and important. The energy and nutrients that fuel life in the forest are broken down here, recycled, and made available for further use. Take time to examine fallen wood and you will see a host of creatures – insects, mosses, lichens, fungi, and ferns – that take advantage of this decay and in turn make a vital contribution to life in the forest by enriching the soil from which forests grow.

LICHENS AND BACTERIA
Both bacteria and lichens, those tangles of algae and fungi, thrive in areas of dappled shade.

CENTIPEDE
A centipede's soft, permeable skin loses moisture, so this tiny predator inhabits damp, sheltered places.

cavity in log becomes home for bees and wasps, even if made by other insects

loose bark offers shelter, food, and moisture

ivy growing over log provides shelter – its berries are a vital food source in winter

INSECT LARVAE
Some fly and beetle larvae inhabit decaying wood; others feed on the fungi associated with it.

LONGHORN BEETLE
Some woodland beetles, such as this longhorn, develop their larvae in dead wood, where they are sheltered and can break down the rotting log into digestible materials.

holly grows in the shelter of oak trees - look for young plants on fallen oak

MONITORING A LOG

A freshly fallen log offers a habitat for wildlife for many years, but the creatures that exploit it change as time goes by. This is a perfect chance for you to watch and record what happens as a fascinating ecosystem develops. Keep a "log file" with notes, lists of wildlife inhabiting it, and photos – especially from a fixed position nearby – over several years.

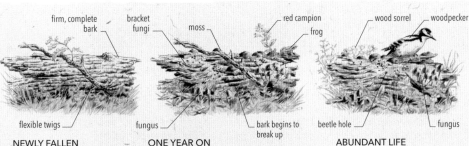

firm, complete bark
bracket fungi
flexible twigs

NEWLY FALLEN
A fresh log with firm bark is a challenge for recyclers. Weak points allow beetle larvae and fungal spores to attack.

moss
red campion
frog
fungus
bark begins to break up

ONE YEAR ON
The bark begins to break up, and twigs become brittle. Mosses, lichens, fungi, and plants appear, and larvae, centipedes, and ants thrive in and around the decaying log.

wood sorrel
woodpecker
beetle hole
fungus

ABUNDANT LIFE
The log breaks up, while plants, mats of moss, lichen, insects, spiders, woodlice, and feeding birds thrive in and around it.

beech leaves cover fallen wood providing extra shelter and food for insects and their larvae

grasses set seed on patches of damp fibre and decaying leaves in open cavities

bore holes show activity of beetle larvae

fungi are found on rotting wood, from which they take nutrients

ferns are common in woodland and many grow on fallen wood

SPIDER WEB
Many spiders drape tangled sheets of web over stumps and logs, rather than creating the usual "cobwebs" across open spaces.

MOSS
The moist, sheltered habitat of a rotting log in woodland allows many moss species to thrive.

Bark life

Bark does far more than just protect the tree, it is also a vital lifeline for many insects, birds, and other wild creatures.

Examining bark

If you look closely at bark you will be able to spot the tell-tale signs of life – from insect trails and bore holes to cavities that provide homes for birds and spiders. Some insects lay their eggs in bark because it provides an insulating layer for larvae over the winter, as well as a good hiding place from predators. Birds, such as treecreepers or woodpeckers, feed on insects in and on bark, or store food caches within it. You may also see a well-camouflaged moth, blending in on a branch. Whether brightly coloured, flaking or polished, scaly, smooth, or woody, bark is an ecosystem in itself.

BARK BEETLE
Lift a flake of loose bark and you may find a host of patterned lines. What looks like an abstract artist's design is in fact the work of beetle larvae, which burrow through the tree's surface.

TELL-TALE HOLES
Look for small holes on a tree trunk – this means a woodpecker has been active – in this case, a white-backed woodpecker is probing for bugs.

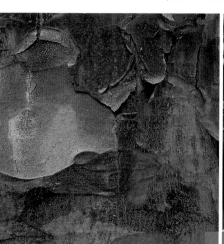
MOTHS
Many moths rest on trees. Some stand out, while others resemble pieces of bark.
BUFF-TIP MOTH

BRACKET FUNGUS
This fungus grows under tree bark as a parasite, taking nutrients from it.

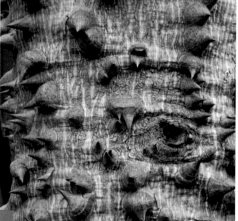

Bark types

Bark differs from tree to tree; on birches it is merely a thin skin, while on conifers it may be as much as 30cm thick. As the tree grows, a thin inner layer, called the cambium, continually produces new bark. The cambium grows either in large sheets, creating peeling or sheath-like bark, or in overlapping arcs, producing a cracked effect. Examine the colours, structures, and patterns and you will soon see how each type of tree has its own distinctive type of bark.

1 The paperbark maple's peeling bark reveals many different layers.

2 Wild cherry's bark forms a ringed or banded pattern.

3 The floss silk tree's bark is thorny, particularly when young.

4 The striped maple derives its common name from its bark.

5 English oak has rough and deeply fissured bark.

6 Sweet chestnut's bark becomes spirally ridged with age.

7 The smooth bark of birch is marked by raised pores (lenticels).

8 Sycamore bark can become grey-brown and flaky with age.

9 Marks and cavities give holly bark a calloused appearance.

10 Beech bark becomes mottled and fissured as it matures.

11 Plane trees have smooth bark that may flake in irregular shapes.

12 Chinese red birch has shiny bark with pronounced lenticels.

BARK RUBBING

You can study different types of bark by taking rubbings with sheets of thick, strong paper and a stick of wax crayon or charcoal. Bark rubbings reveal different patterns without the distractions of surface colours, moss, or other debris. Make your own, carefully labelled collection to highlight the variety of trees you have discovered in a woodland in your local area.

BARK RUBBING

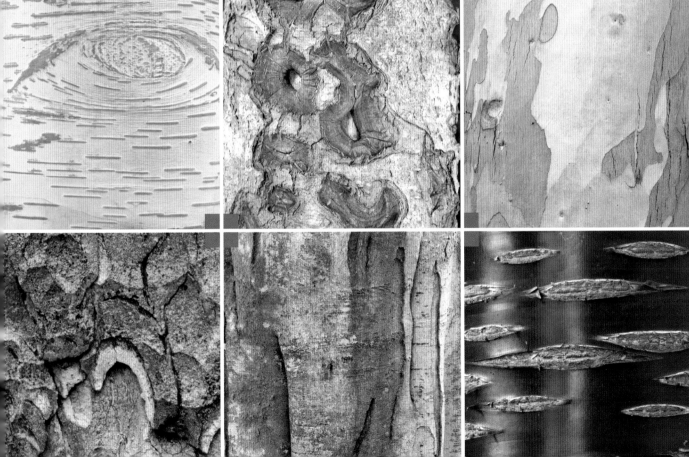

Looking up

Most of us forget to look up, but glance skywards in a forest and you will discover another layer of woodland activity to explore.

Pollination

Many trees can produce an identical version of themselves by sending out suckers. Flowers, on the other hand, exchange genes through pollination with another plant of the same species, creating genetically different offspring. Pollination is carried out by insects, such as bees or mosquitoes, by birds carrying pollen from flower to flower, or by wind. Look up in any woodland and you will often be able to see pollination in action.

SINGLE SEX
Male and female flowers appear on separate bay laurel, or sweet bay, plants. Fruit only grows on female plants, as here.

MOSQUITO

MIXED BLOOMS
Both male and female flowers are found on each alder buckthorn plant, which are pollinated by mosquitoes.

Fruits of the forest

Trees are static and rely on external factors to help disperse their seeds. Forests are filled with all kinds of seeds, nuts, berries, and fruits that feed all sorts of animals, from the smallest birds to the largest bears. These animals aid the plants by transporting seeds in their digestive tract and depositing them at a different location. Jays and squirrels bury acorns for later use, thus helping to spread oak woodland.

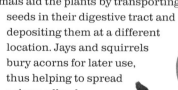

BLACK CHERRY

WALNUT

Never eat wild nuts or berries as they may be toxic.

SWEET TREAT
Juicy berries are packed with high energy and sugary nutrients to make them irresistible to bears.

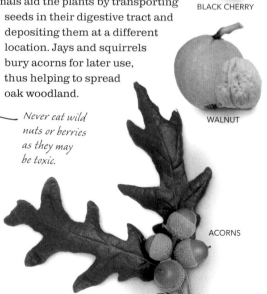

ACORNS

Galls and miners

Close inspection of some woodland plants may reveal some odd-shaped growths. These galls are produced in response to parasites, such as fungi, bacteria, insects, and mites. Some insects use galls to provide food and protection for their larvae. The larvae of leaf miners create tunnels by feeding on the cells between the upper and lower surfaces of leaves.

GALL WASPS
Some wasps, such as this marble gall wasp, produce galls on oak trees. Gall wasps are tiny with usually shiny red-brown or black bodies.

oak spangle gall
— eggs laid by wasps

SPANGLE AND CHERRY GALLS
Most galls are produced by wasps - in this case creating red galls on the underside of a leaf.

"mines" show where leaf miners have tunnelled

LEAF MINERS
Some caterpillars, such as the horse chestnut leaf miner, injure leaves, but cause no lasting damage to trees.

FALLEN WONDERS

If you want to find out more about the variety of invertebrates that live in a healthy tree or shrub, carefully place a white sheet on the ground or hold out a rigid white board beneath it. Then, gently shake the branches above it to encourage insects, spiders, mites, and any other tiny animals to let go and fall on to the sheet or board. Beware of any fallen branches that may be dislodged by your movements. Identify what you can, record everything you find, and then leave the animals to return to their habitat.

LOOK AND LISTEN
Life in the canopy is best observed by lying down. Take time to relax, watch the action, and listen to birdsong and the hum of insects.

Forest birds

Birds are some of the most lively, colourful, and noisy forest inhabitants, yet they are apt to fly away at the least disturbance.

Woodland chorus

Getting to grips with woodland birds is difficult because many hide in the foliage, and some become very quiet in the summer. To get a better idea of what lives in a wood, and for the sheer enjoyment of hearing the birds at their best, try to hear a spring dawn chorus (see panel, below). Birds sing or call for a variety of reasons – to defend territory, ward off rivals, find a mate, warn of a predator, or locate their chicks. Over time you can learn to distinguish not only the calls of different species but also the types of calls.

RECORDING BIRD SONG

Distinguishing bird song can be difficult but rewarding. Listen to birdsong CDs to prepare for when you are out in the field. Many field guides transcribe bird sounds into words such as *tiks*, *chaks*, and *tchuks*, which are very useful for identification. Try making your own notes when you're out and about – be creative with your descriptions.

Chshree-ip
Schrree-eew
Shrr-ooo

SOUND GUIDE
When transcribing bird sounds, the lines above the word are used to show variation in pitch.

SWEET SONG
In spring, male common redstarts sing a sweet, high-pitched song from first light. This is a declaration of territory.

THE DAWN CHORUS
This avian choir is at its best in spring, when males attract females and warn off rivals, but you will need to get up early to catch it – as early as 4am in some areas. Try to choose your location the evening before, as birds such as blackbirds often sing from the same song-post they settle on at dusk. Sit quietly, don't wear bright colours, and enjoy the performance.

Woodland birds

PIED FLYCATCHER
Summer visitors to European woodlands from Africa, pied flycatchers nest in tree holes or nest boxes.

JACKDAW
The jackdaw is a noisy crow with pale eye, black cap, and grey hood. It is widespread in most of the northern hemisphere, where it is often found in flocks around old buildings, woods, or cliffs.

HAWFINCH
A secretive bird, the hawfinch has a bill strong enough to crack open cherry kernels.

Nesting

Birds' nests are not long-term "homes" – they are solely for hatching eggs and rearing chicks. We can learn much about birds from their nests, but take care to never disturb an active one. Over time, the average number of eggs, number of clutches, and chick survival rates give us vital data on the health of bird populations.

SONG THRUSH NEST

REDSTART NEST

BUILDING MATERIAL

Nests vary from mere scrapes in bare earth to extraordinary constructions, often with tough outer structures of twigs (top), roots, and grasses, lined with softer hair, moss, fur, or feathers (above, and left).

Migrants

Migration is one of nature's extraordinary events. In the northern hemisphere, birds head north in spring, to exploit a temporary glut of food, and return south in autumn, often to share winter quarters with similar species that stay there all year. Some, such as geese, travel in families and learn routes; others migrate alone, like cuckoos, navigating by instinct (see p.15).

SUMMER MIGRANT

The wood warbler is a visitor to broadleaved woods in Britain and Europe during the summer. It winters in tropical Africa.

DEFIANT SONG

The robin, found widely throughout Europe, uses its elaborate song to defend a territory by warning off rival males.

WOODPIGEON

This is a big, colourful, bold, and abundant European woodland bird. It tends to be shy in farmland, where it is shot as a pest; it is often tame in town parks.

CHAFFINCH

A colourful European finch that is found in woodland, parks, and gardens. The female has the same white wing patches, but is much less pink than the male shown here.

CHIFFCHAFF

This small European warbler is typically rather plain, but can be identified by its *chiff chaff chaff chiff chaff* song.

Deciduous close-up

Deciduous woodlands around the world harbour an enormous diversity of life – including many flowering plants, mammals, insects, and birds – that varies according to its range and with the seasons. Get to know your local forest well, month by month.

Forest flowers bloom in early spring before the canopy shuts out light.

DOG VIOLET

BLUEBELL

WOOD ANEMONE

BUGLE

LESSER CELANDINE

FORGET-ME-NOT

Look out for mammal bones on the forest floor.

SQUIRREL SKELETON

Insects may be found sheltering on fallen wood.

CENTIPEDE

SWALLOW-TAILED MOTH

Nuts and seeds ripen and fall in autumn.

RED MAPLE WINGS

SYCAMORE KEY

SWEET CHESTNUT

Leaves of trees and shrubs provide shade in summer and carpet the floor in autumn.

BIRCH LEAVES

PAPER BIRCH LEAVES AND CATKIN

HORSE CHESTNUT

BLACK BRYONY

OAK LEAVES

ORANGE PEEL FUNGUS

MOREL FUNGUS

HAZELNUTS

ACORNS

Many fungi live in close association with trees and are found at their roots.

FLY AGARIC

COMMON FROG

Look for amphibians near forest pools or decaying wood.

FIRE SALAMANDER

HONEY FUNGUS

The forest year

Few places reflect the changing seasons as well as a deciduous forest – its colours, sounds, and scents reveal the natural cycle of life.

Observing the changes

Appreciate the uniqueness of each season in a wood by keeping all your senses alert. Spring is the best time for listening to the birds, and the lush growth of summer provides great opportunities for plant hunting. Autumn brings with it the scent of rotting leaves and fungi, while the peacefulness of a winter wood should never be underestimated.

BRINGING UP BABY
Summer is the time when mammals are busy rearing young. Fox cubs may be seen playing outside dens.

spring flowers provide autumn fruit

CHERRY BLOSSOM

1 Spring stimulates new life. Longer days allow the increasing energy from a hi gher sun to pour through the trees. Woodland flowers thrive in the light, before the growing canopy casts deep shade.

2 Summer is a quieter time as animals and birds move on from the frenzy of courtship and defining territories to the hard grind of raising families. At this time of year, ferns and lichens become more obvious than flowers.

Examine plants to see insects feeding on them – many are camouflaged.

EATING GREENS
Caterpillars gorge themselves on lush, summer foliage.

ROOTING FOR FOOD
Wild boars turn over soil in their search for roots and invertebrates.

NEW SHOOTS
Conditions in spring – longer, warmer days with more sunlight – give plants, such as this hazel, the energy to put forth new growth.

SPRING TOADS

Common toads spend the winter buried in soft ground and emerge early in the year. They gather in shallow water in March, where females lay long strings of jelly containing three or four rows of black eggs - quite unlike the shapeless mass produced by frogs so they are easy to tell apart. Toad tadpoles often form dense shoals, have flatter bodies, and are blacker than frog tadpoles.

TRACK A FOREST YEAR

Really get to know a particular wooded area, or even a single large tree, by visiting and recording it throughout the year. Choose somewhere close by and easy to get to so that you can visit every week. Note down anything that is new or has changed, and take photographs to compare at the end of the year. Identify and count all the trees in your chosen area, and as other plants begin to appear identify what they are and note their flowering dates. Listen out for birdsong, try to identify it and record it on a sound recorder if you can. Look closely for insects, fungi, and other wildlife of the wood.

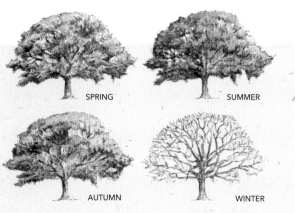

SPRING

SUMMER

AUTUMN

WINTER

earwigs burrow into the ground to survive winter

EARWIG

OUT IN THE COLD
Some mammals, such as badgers, stay active all winter.

3 Autumn sees a lowering sun, and with it, reduced light penetrating the forest. Insects decline and migrant birds leave, but the immense bounty of nuts, seeds, and berries that remains tempts some species to stay and begin storing food for the winter.

AUTUMN FEAST
Rodents such as this yellow-necked mouse bulk up on berries, nuts, and insects in autumn before food becomes scarce in winter.

Look for butterflies and moths in bushes and hiding on tree trunks.

FATAL FREEZE
As the weather turns cold some moths and butterflies hibernate while others die.

4 To survive winter, some animals hibernate, while birds roam in mixed, nomadic parties for safety. A lack of foliage can make birds and animals easier to spot at this time of year, and look out for their tracks in mud and snow.

WINTER GREEN
Look up to see mistletoe clinging to bare winter trees.

Signs of life

Mammals are sometimes difficult to see in a forest, but finding evidence of their activities – dens, nests, and tracks – is often much easier.

Making tracks

Most tracks are left in mud, which can last a few days, or snow, which can be very short-lived. The best prints are those found in mud, as it preserves details of the structure of the foot or paw. Snow tracks are far less well-shaped, unless they have been made in a thin layer of snow on soft ground, especially after fresh snowfall. Look for prints and tracks around muddy puddles, on wet trails in the wood, or near rivers and streams.

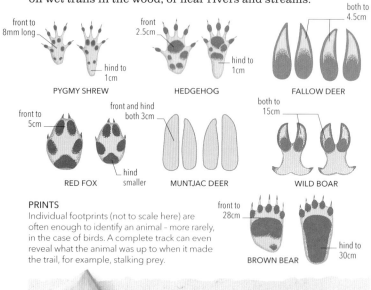

front to 8mm long · hind to 1cm
PYGMY SHREW

front 2.5cm · hind to 1cm
HEDGEHOG

both to 4.5cm
FALLOW DEER

front to 5cm · front and hind both 3cm · hind smaller
RED FOX

MUNTJAC DEER

both to 15cm
WILD BOAR

front to 28cm · hind to 30cm
BROWN BEAR

PRINTS
Individual footprints (not to scale here) are often enough to identify an animal - more rarely, in the case of birds. A complete track can even reveal what the animal was up to when it made the trail, for example, stalking prey.

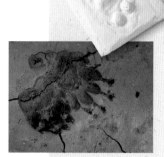

MAKING A PLASTER CAST

Good tracks are worth preserving. Photographs are valuable, but it's rewarding - and sometimes more informative - to make a plaster cast. First, spray the print with very light oil, such as cooking oil, then surround it with a "wall" made up of thin strips of wood, plastic, or thick card, so that the plaster has a barrier and won't seep away. Carefully pour a runny plaster mix over the print. When it is set hard, lift it to reveal a "negative" cast. You can then make a second cast of the first to get back to the original imprint.

BADGER TRACK AND CAST
This broad, long-clawed print of a badger's foot has been preserved for future study in a plaster of Paris cast.

Resting places

Most birds build nests for hatching eggs and raising their young, but mammals use their resting places as shelter for much longer. Some, such as a badger's sett, are daily retreats; others, like a bear's den, are seasonal. Shelters can be found in all sorts of places, some are dug into the ground, while others might be nestled in a tree.

1 Look out for a loosely structured, roofed nest – this will be a squirrel's drey. A summer drey is used for resting, while a winter, or family, drey is stronger to provide better shelter.

2 This beech marten is looking out of a tree hole. This forest-dweller uses natural holes in trees or rocks rather than digging its own or using those abandoned by other animals.

3 Fallow deer don't make dens, but leave their well-camouflaged fawns hidden in sheltered hollows where they keep still to avoid detection.

4 Foxes dig burrows called "earths". Simple ones are daytime or emergency retreats, while larger ones are for raising young.

5 In winter, hedgehogs may hibernate under the roots of trees, in sheds, under timber buildings, or in old rabbit burrows like this.

6 A badger sett looks like a large rabbit hole, but for the coarse grey hairs, tracks, piles of old bedding, and latrines.

Watching at night

Many animals are active at night, so an evening watch can yield amazing sights. You may be in for a long wait, so get comfortable. Nocturnal animals have poor vision, but hear and smell brilliantly, so sit still on a bank or tree above their level and downwind from them – keep quiet and don't smoke or wear perfume. Use a torch with a red filter as red light is at a wavelength that animals can't easily see.

Tips for your hide
1. Prepare a comfortable perch in advance.
2. Wear dull, dark clothes.
3. Turn off mobile phones.
4. Cover your torch with red cellophane.
5. Always tell someone where you are going.

BADGER TRACK Wait quietly and don't make a sound at a sett and you may get a close view.

Pine plantation

Demand for timber has led to extensive planting of conifers in stands of single species, such as sitka spruce. Young plantations are dense and dark, but thinning as they age creates a more natural condition. Before they are harvested, trees may lose branches or be blown over by gales. This creates space for wildlife, where fungi, such as morels may grow.

SITKA SPRUCE

MOREL FUNGUS

Coniferous forests

Conifer forests are not all dark swathes of "Christmas trees". While many plantations are poor habitats for wildlife, natural forests are home to a diverse range of plants and animals, many of them found nowhere else.

Montane conifers

Mountain pines, silver firs, Norway spruces, and yews are all examples of conifers that grow at high altitudes, often up to the tree line, in the Alps, Pyrenees, and other mountain ranges. These trees are able to cope with snow and hard frosts for much of the year. Among the resident birds are capercaillies and crossbills.

YEW

CAPERCAILLIE

Scots pine forest

Scots pine forest grows from Scotland eastward across Siberia and stretches as far south as the Mediterranean on high mountains. The Scots pine was the only northern European pine to survive the last ice age and it has a fragmented distribution. Mature forests are home to predators, such as wildcats, and support rich plantlife, including heathers and cowberries.

COWBERRY

WILDCAT

Taiga

Taiga is the cold forest zone south of the Arctic tundra. Trees, such as the black spruce, have shallow roots to exploit the thin soil, downward-pointing branches to help shed snow, and dark needles to absorb weak sunlight efficiently. Ground beetles shelter in needle litter while wolverines have thick fur coats for insulation.

GROUND BEETLE

WOLVERINE

BLACK SPRUCE

BILBERRY

COMMON
BUZZARD FEATHER

*Keep quiet and scan the canopy
for shy animals such as squirrels
and martens – watch for any
movement against the bright sky.*

Native pinewoods

**With their year-round greenery, pinewoods have a
characteristic beauty. There is a wealth of wildlife
to discover – once you know where to look.**

A walk through pinewoods can be a
satisfying experience that engages
all your senses. Experiencing the
fresh scent of these evergreen forests
is a distinctive part of any visit to this
habitat, and like all forests, it can be

a quiet, undisturbed place for a
budding naturalist to explore. From
insects scurrying on the forest
floor to birds calling in the canopy,
pinewoods harbour a wide range of
animals, as well as plants and fungi.

*Search the
undergrowth of
open glades
for colourful
herbaceous plants,
such as ling and
other heathers.*

MOSS

LADY'S-TRESSES
ORCHID

Listen for bird calls and other tell-tale noises. The sound of pine cones falling may indicate squirrels or birds, such as crossbills, feeding above.

LING

LADYBIRD

WOOD HEDGEHOG MUSHROOM

Remember to keep one eye on the ground: here you might find ant nests, flowers, or fungi, or the entrances to badger setts and fox earths.

WOOD ANTS

STRANGE FRUIT
Many fungi are poisonous and it is important to never pick or eat any mushrooms you find in the wild.

Forest fungi

Looking for fungi adds to any walk. You can find mushrooms and toadstools in a variety of shapes – but that's only one part of the story.

What are fungi?

Fungi are neither plants nor animals, but organisms that feed on rotting material, breaking it down to enrich the soil. The parts we see – from delicate toadstools to thick, chunky brackets – are only those involved in reproduction and are known as fruiting bodies. Underground, thread-like filaments known as *hyphae* spread out to form a colony-like *mycelium* – a mass often many hectares in size. Some *mycelia* are thousands of years old.

MOULDY WOOD
Look out for slime mould, which closely resembles fungi, growing on damp, decomposing wood.

Fruitbody shapes

Most fungi have a stem topped by a cap. In puffballs, though, stems are almost invisible, while in others, such as stinkhorns, the stems are more striking than the caps.

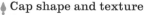

Cap shape and texture

Caps may be conical, domed, flattened, or dish-shaped and may open from round "buttons" into broad "dishes". They can be dry and flaky, silky, or greasy.

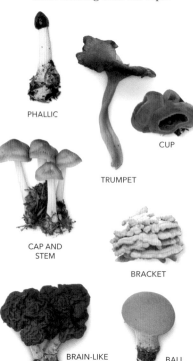

PHALLIC

TRUMPET

CUP

CAP AND STEM

BRACKET

BRAIN-LIKE

BALL

CONVEX

FUNNEL

GROOVED

DEPRESSED

CONICAL

LOOSE SCALES

A BENEFICIAL RELATIONSHIP

Many fungi live in close association with plants and algae, usually to the benefit of both; scientists estimate that more than 90 per cent of plants need fungi for their own survival. Fungi help plants take up nutrients, such as nitrates and phosphates from poor soils, in a system known as "mycorrhizal symbiosis" – a partnership that benefits both parties. Fungi living in close association with algae form lichens, which are abundant in unpolluted forests, as well as on rocks, roofs, and walls. Some fungi parasitize animals, such as bees, while some ants and beetles cultivate fungi for food.

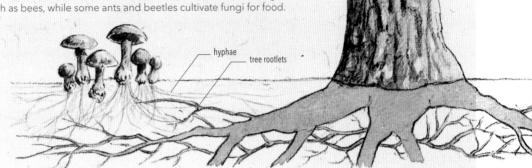

hyphae

tree rootlets

FUNGUS ROOTS
The word *mycorrhizae* means "fungus roots", and these are what colonize plant roots, helping the fungi to access carbohydrates and the plant to absorb water and mineral nutrients.

Stem

The stem, called a stipe, raises the cap to allow a fungus to release spores to reproduce. In some species, the cap bursts open from a spherical shape, leaving a lower ring on the stipe.

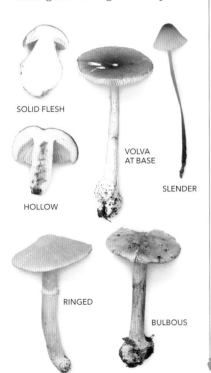

SOLID FLESH

VOLVA AT BASE

SLENDER

HOLLOW

RINGED

BULBOUS

Colour

Colour is important in identification but subtle shades of brown, pink, yellow, and orange can be difficult to describe. Find a colour chart, or use your own terms for comparison.

RINGS OF COLOUR

DARK CENTRE

SPORE PRINTS

Spores are ejected from a fungus by a build-up of internal pressure or by the force of a raindrop. You can easily take a spore print to help identify a fungus. Put the cap on paper, place a glass over it, and leave it overnight. Then carefully remove the glass and fungus, and quickly spray the image lightly with hairspray to "fix" the print.

Gills

Some fungi have caps with fine plates or "gills" underneath – this is where the spores are produced. Learning the structure of the gills will help you identify the species of fungus.

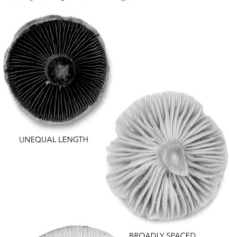

UNEQUAL LENGTH

BROADLY SPACED

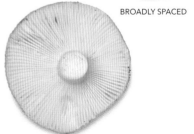

CROWDED

Coniferous close-up

Coniferous forests are widespread in the northern hemisphere and support an array of specialized flora and fauna throughout their range. The exact species of fungi, insects, and plants will depend on location, but all are home to a rich variety of wildlife.

PUFFBALL FUNGI

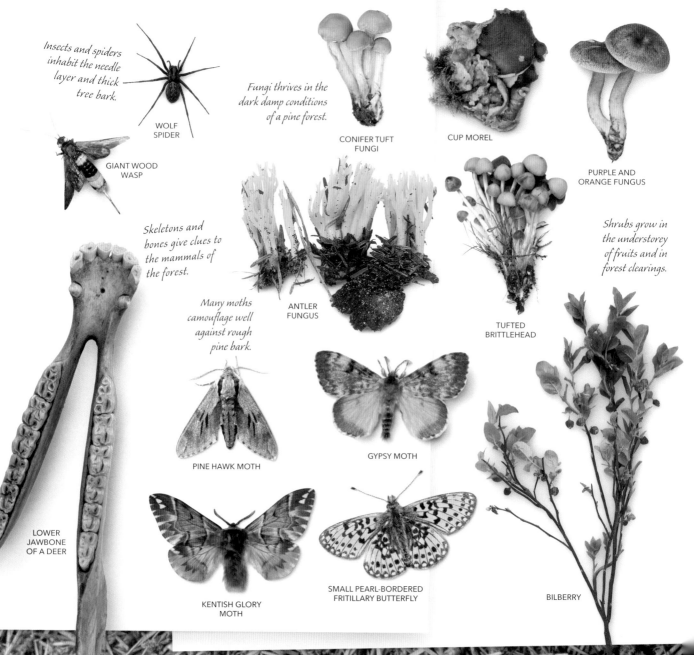

Insects and spiders inhabit the needle layer and thick tree bark.

WOLF SPIDER

GIANT WOOD WASP

Fungi thrives in the dark damp conditions of a pine forest.

CONIFER TUFT FUNGI

CUP MOREL

PURPLE AND ORANGE FUNGUS

Skeletons and bones give clues to the mammals of the forest.

ANTLER FUNGUS

TUFTED BRITTLEHEAD

Shrubs grow in the understorey of fruits and in forest clearings.

Many moths camouflage well against rough pine bark.

PINE HAWK MOTH

GYPSY MOTH

LOWER JAWBONE OF A DEER

KENTISH GLORY MOTH

SMALL PEARL-BORDERED FRITILLARY BUTTERFLY

BILBERRY

WINTERGREEN

SITKA SPRUCE
NEEDLES AND CONE

*Lichens hang
from low branches
and spread over
dead wood.*

JUNIPER

WILD
RASPBERRIES

FOLIOSE
LICHEN

BELL
HEATHER

NORWAY
SPRUCE
CONES

SCOTS PINE CONES

*Cones litter
the forest floor.*

Coniferous specialists

Cones solve problems for trees in tough conditions but can prove tricky for animals determined to eat their seeds.

TOP OF CONE

tightly packed scales at base

BOTTOM OF CONE

The pine cone

Most pine trees bear both male and female cones. Male cones are small, with modified scales covering pollen sacs. Female cones are the more familiar large, woody cones that contain ovules that, after being fertilized by pollen, develop into seeds. While their structure is similar, you'll find the size, shape, and woodiness of cones vary from species to species.

cone tip

You can tell if humidity is high or low by whether a cone opens or not.

closed scale

open scales release seeds

algae growing on surface

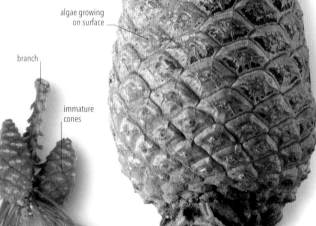

branch

immature cones

SCOTS PINE CONES
Rosy-pink female flowers turn purple in summer, with small scales that become bright green but woody the following year. The year after that, the cones are mature, and turn a dull grey.

clustered pine needles

attachment

CLOSED CONE
Female cones have seed scales, which open initially to receive pollen, then close tightly while the seeds mature. Later, they will also close in wet weather to protect and retain the growing seeds.

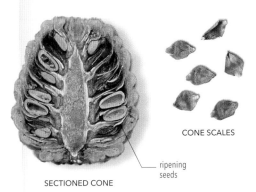

CONE SCALES

SECTIONED CONE

ripening seeds

Start a cone collection and look for clues to find out which animals opened them.

protrusion on scale

OPEN CONE
Mature cones open in dry weather – with reduced moisture content – to ensure seeds are released in ideal conditions for wind dispersal.

Cone crackers

Seeds within pine cones lie deep between the scales at the base of thin, flat, wing-like structures. They are nutritious, but difficult to reach, and eating a whole cone, with its hard, rough, sharp-edged, woody scales, is inefficient, so many animals have developed ways to get inside. In damp weather, scales close up tight – dry conditions open them – but this happens many times, even long after the seeds have been dispersed and long after the cone has fallen to the ground. Seed-eaters must first decide which cones are worth their attention.

1 Crossbills, such as this two-barred crossbill, have evolved into many species, often in response to the size and shape of particular cones. They push their mandibles between scales, then close or twist their bills to open them. Seeds are then extracted with the tip of the tongue.

2 Squirrels simply bite the scales and gnaw their way in to the seeds. You can easily tell cones bitten down by a squirrel from those worked on by a crossbill.

3 Woodpeckers often wedge cones into bark to make it easier to peck between the scales and extract the delicate seeds.

WOODPECKING

Woodpeckers chisel into living or dead wood – their long, sticky, spiny tongues, which wrap around the skull, probe deeply into holes in order to extract insect larvae. A woodpecker steadies itself by using its stiff tail as a prop, and grasps the tree firmly with specially adapted feet – two toes point forward, two point back. It also has sinewy attachments at the base of its bill and around its brain to reduce the shock of the fierce bombardment of bill on wood.

tongue   brain

CLINGING TO TREE

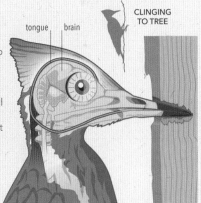

In a rut

Red deer make great wildlife watching, and never more so than in autumn, when stags fight for the right to mate.

Originally forest mammals, red deer are now also found on open mountains and moorland. They are large animals – big males (stags) are more than 2m long and 1m high at the shoulder. The male red deer's bony antlers are shed annually. These have several points, or "tines", and each year's antlers are bigger and more complex than those from the year before. Their growth requires a massive investment of energy, but, in winter, when food is short, the antler is simply "dead bone", making little demand on the deer. Stags are in their prime at about eight years old; younger and older ones rarely secure a harem of females, or hinds. In autumn, at the onset of the mating season (rut), males are very aggressive and if you see them you should keep your distance. You can hear stags roar almost constantly, and see them pacing together to assess each other's size and condition. Weaker males back down, but those of equal strength may fight. Females are attracted to the males who best demonstrate their fitness, which can be expressed by loud, frequent roaring.

AN ANTLER'S YEAR

A deer's antlers develop through the seasons. In spring a small, sensitive knob, or "pedicle", gives rise to the new antler; a covering of soft skin called velvet provides blood and oxygen, but gradually shreds and falls away when the blood supply to the antler ceases in autumn. The antler calcifies, becoming hard; then, at the end of winter, it falls away. Look out for shed antlers on the ground in spring or velvet in late summer.

pedicle

growing antlers in velvet

full grown, velvet slowly being shed

SPRING

MIDSUMMER

LATE JULY

HEAD TO HEAD
When two stags are evenly matched, they lock antlers and start a serious shoving match. Occasionally, a sharp antler point will cause damage, but this is rarely a fatal wound.

Elusive creatures

Red deer are so spectacular you can watch them from a distance, but most forest animals are shy and hard to find. In fact, finding some of them could take a lifetime.

PINE MARTEN
Shy pine martens are most likely to be viewed at nature reserves in Scotland and northern Europe.

FLYING SQUIRREL
This largely nocturnal animal of Estonian and Finnish forests is very secretive.

GREY WOLF
The largest wolf is now scarce in its much-reduced European range.

LYNX
All four species of lynx, including the European lynx (above), are very hard to find.

117

Scrubland and heath

Most scrublands and heathlands are not naturally permanent habitats – without the maintenance of grazing or fire they would soon become woodland. Exceptions are those where altitude, nutrient poverty, or wind prevent the climax community from succeeding. These are open, hot, and dynamic places where communities "boom and bust" in rapidly changing conditions, and in the modern world they are sadly often endangered by human activities such as clearing for agriculture or fragmentation through development. For this reason they are frequently rare, their species component exotic and exciting – and naturalists love them!

Lowland heath

Originally woodland cleared by prehistoric humans, lowland heath is found mainly in northwestern Europe. It is defined by short plants and shrubs, such as heather, gorse, and types of grasses, and is home to a range of specialized animals, including birds such as the woodlark and nightjar. When left to its own devices, without being controlled by grazing and fire, lowland heath reverts to woodland.

YOUNG NIGHTJAR

BELL HEATHER

Scrublands and heaths

Heaths and scrublands are habitats in transition. They are usually found on nutrient-poor soils and are shaped by factors such as grazing and fire. Their unique conditions provide a home to many interesting types of wildlife.

Moorland

Moorland is wetter and colder than lowland heath, and it occurs at higher elevations, usually 300m or more above sea level. It also has fewer shrubs than lowland heath, and plants such as heather are more prominent, alongside other low-growing species such as bilberry and bog myrtle. Upland birds, such as grouse, thrive here, as do birds of prey, and reptiles such as adders.

ADDER
(COMMON VIPER)

BOG MYRTLE

Garrigue

Found around the Mediterranean, garrigue and maquis are two closely related habitats. Many of the same plants and animals are found in both. Garrigue has developed in areas where the underlying rock is limestone and the soil is thin and stony. Low-growing, drought-resistant plants, such as rosemary, dominate the landscape, along with pine and cork oak trees. Lizards and snakes are common, as are mammals such as wild boar and deer.

OCELLATED LIZARD

ROSEMARY

SARDINIAN WARBLER

Maquis

Unlike garrigue, which is a relatively open habitat, maquis is dominated by small trees and large shrubs that form dense, often impassable, scrub. Vegetation typically grows 3–5m tall, with dwarf oak, carob, olive, fig, and almond trees, climbers such as ivy and spiny asparagus, and shrubs such as strawberry tree, broom, bay laurel, and myrtle. Formerly forest that was cleared for grazing land, firewood, timber, and charcoal production, maquis provides a home for wild boar and deer, birds such as Sardinian and Dartford warblers, and reptiles including Hermann's tortoises, lizards, and snakes.

OLIVE TREE

Life in the scrub

Scrubland creatures are engaged in a constant battle to survive. Small insects are preyed on by larger ones, which in turn are eaten by birds, reptiles, and mammals.

TAWNY
MINING BEE

Home defence

Insects occupy all levels of the heathland and scrubland habitats, and competition for a desirable spot can be intense. Look for territorial tussles between beetles, and for male butterflies chasing away intruders. Also keep an eye out for insect-hunting birds – warblers move through the scrub, searching for spiders and caterpillars, while shrikes perch prominently, scanning for large beetles and lizards.

SPIDER TACTICS
While some spiders hunt their prey by chasing it, others use webs or lie in wait in holes, jumping out to grab passing insects.

MEGA PREDATOR
Spider-hunting wasps are top predatory insects. They can even overcome a large spider with their deadly sting.

SOUNDS OF THE SCRUB
In the heat of summer, an orchestra of crickets, grasshoppers, and cicadas produce a constant trilling, chirping, and rasping. Look for cicadas on tree trunks.

Underground life

From ants to scorpions, a huge variety of small creatures use underground burrows for safety and to breed in. They either excavate their own accommodation or take over a home that originally belonged to something else. Watch holes in the ground carefully to see what species emerge. Evidence of mining bees is easy to spot – look for tiny, volcano-shaped mounds of spoil in areas of exposed soil. The dirt is carried up by the bees as they excavate their chambers, and they leave it piled around the shaft entrance.

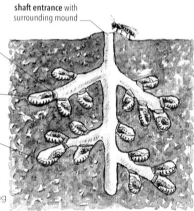

shaft entrance with surrounding mound

chamber seal is closed by female with wax

side chamber contains small larva feeding on stored pollen

small larva develops into pupa in chamber

LEFT ALONE
Each mining bee nest is dug by a single female that seals the entrance and dies, leaving the young to develop alone.

Rearing moths and butterflies

Search for caterpillars on plants such as sages and brambles, then transfer them to a secure but ventilated transparent container. Put a little soil and dead plant material at the bottom, and each day provide fresh supplies of the same type of leaves you found them on. Eventually, the caterpillars transform into cocoons. Release the adults when they emerge after a few weeks (longer with some species).

EMPEROR MOTH
Emperor moths are easy to rear from caterpillars, but live only a week or two until they find a mate, breed, and die.

PHEROMONE TRAP

Some moth species locate each other using chemical signals (pheromones), including those given off by females to attract males. Watch this behaviour by rearing moths such as emperors. Place a newly emerged female in a pheromone trap made of plastic or netting, folded or stitched into a tent-like shape and suspended from a branch. Watch males arrive and whirl around as they try to reach her.

ATTACK FROM ABOVE
Burrowing solitary wasps can be found worldwide. They swoop down and paralyse their prey before carrying it off to their nests as food for their larvae.

Scrublands close-up

Dry, scrubland habitat is home to an array of hardy plantlife, insects, and mammals. The actual species will vary, depending on where you are in the world.

GREEN HAIRSTREAK

Insects occupy all levels of scrublands — in winter, look for moth cocoons on the ground.

MOTH COCOON

CYTISUS BROOM

SCARCE SWALLOWTAIL

BEAUTIFUL YELLOW UNDERWING

PRAYING MANTIS

Short plants and shrubs, including herbs, characterize the transient scrublands.

FALSE DEATH CAP MUSHROOM

PENNY BUN MUSHROOM

YELLOW-GIRDLED WEB CAP MUSHROOM

OLIVE TREE LEAVES

Search through areas of dense shrubs and plants, which provide shelter for fungi species.

FLAT MUSHROOM

PELARGONIUM WEBCAP MUSHROOM

CRAB RUSSULA MUSHROOM

BLACKTHORN

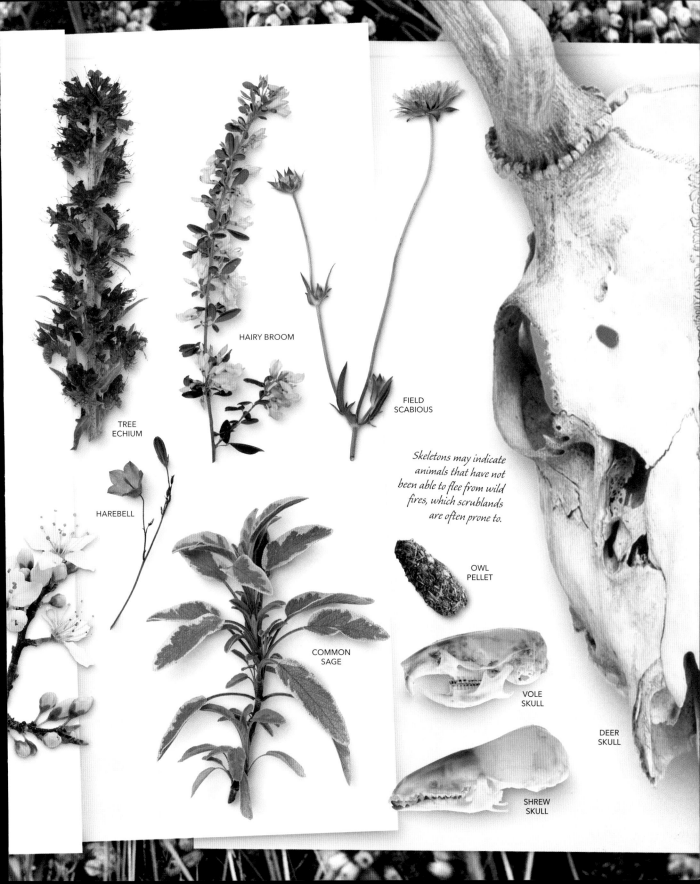

TREE
ECHIUM

HAIRY BROOM

FIELD
SCABIOUS

HAREBELL

COMMON
SAGE

*Skeletons may indicate
animals that have not
been able to flee from wild
fires, which scrublands
are often prone to.*

OWL
PELLET

VOLE
SKULL

DEER
SKULL

SHREW
SKULL

Heathland

The open, exposed character of lowland heath means that much of the wildlife lives on or near the ground.

A changing habitat

Lowland heath is a constantly evolving habitat found in Europe. If it is left unmanaged by humans and unaffected by natural events, such as fire, the open heather habitat is rapidly replaced by scrub, which in turn becomes woodland. In the past, heathland was valued for what it produced, and managed by grazing and the cutting back of invasive gorse and birch. After decades of decline, such activities are being used again today to keep lowland heath in the condition required by the specialized wildlife that lives there.

A VERSATILE PLANT
People in Europe traditionally used heather to stuff pillows and mattresses, for insulation, roofing material and making brooms, as well as in the production of ale and honey. Older plants were burned periodically to promote growth.

HEATHLAND POOLS

Wetland areas are home to species that otherwise could not survive in the heathland environment. Find a suitable pool and try pond dipping for amphibians such as frogs, toads, and newts, and look out for aquatic invertebrates, such as raft spiders. You can also find damp-loving plants, such as bog bean, bog asphodel, and sundew, growing at the water's edge, and see dragonflies and damselflies hunting overhead.

RAFT SPIDER

SOUTHERN HAWKER

BOG BEAN

COMMON TOAD

SPECIALIST SONGSTER
Dartford warblers are found in gorse scrub. In spring, look for males on a prominent sprig as they sing their buzzy, chattering song.

BUTTERFLY BEHAVIOUR
Silver-studded blue butterflies can be seen flying low over heather and feeding on its nectar.

stems were dried and bound together for thatching and broom-making

flowering heads were used like hops to make ale

heather was considered by some to be lucky, and dried sprigs were bunched and sent for sale in the cities

MOVING IN
Thorny gorse readily colonizes open areas. The young plants soon develop into thick patches of scrub and eventually crowd out the heather.

NEW SHOOTS
Heather seeds can lie dormant for decades, but they start germinating as soon as the scrub above them is cleared and light floods in.

SUN LOVERS
Lizards bask in sunny patches of soil between heather. Find a sheltered location and wait quietly for them to emerge.

FINDING SNAKES
Heathland is the best place to spot smooth snakes. These non-venomous snakes are native to southern England, where they are one of the rarest reptile species.

ROOTS FOR FUEL
Slabs of matted heather roots, known as turves, are traditionally dug and burned for fuel during the winter.

HEATHLAND KILLER
Ground beetles, such as the green tiger beetle, can be spotted chasing down prey on areas of exposed soil.

heather has tightly bunched flowerheads

BLOOMING HEATHER
Heather flowers provide nectar for bees, butterflies, and moths, while the leaves are eaten by caterpillars, birds, and livestock.

CORK OAK
LEAVES

CLEOPATRA BUTTERFLY

SPANISH FESTOON
BUTTERFLY

*Find a sunny spot
where you can
watch butterflies as
they fly around
searching for flowers to
extract nectar from.*

Garrigue walk

The scent of aromatic plants and the trilling sound of insects are the essence of the Mediterranean garrigue, especially in the heat of high summer.

The garrigue is a habitat full of subtle sounds. Muted birdsong and the scuttling noises of rodents and lizards are clues to the diversity of wildlife that lives here. Insects are particularly abundant, and sun-loving species seek out open areas of bare soil and exposed rocks. These are also good places to search for lizards and other reptiles.

*Listen for the rustle
of reptiles, including
tortoises as they bulldoze
their way through the
vegetation in search of
plants to eat.*

MYRTLE

STRAWBERRY
TREE

LAVENDER

Where shrubs such as myrtle and strawberry trees start to dominate, the open character of the garrigue gives way to maquis.

ASCALAPHID

TONGUE ORCHID

Summer is a great time for plant hunting, with many types of heather coming into bloom. Look out for their flowers, which can be seen in varied shades of purple, pink, and red.

Grassland

The acacia-freckled African savanna, the silver seas of the high
Andean páramo, or the rolling, flower-rich downlands of southern
England all have an enduring appeal – perhaps because the first
is our own species' primal home. They are all open and warm
habitats that are highly productive, supporting a large number
of herbivores, and in turn, a range of carnivores, from lions and
maned wolves to foxes. Meanwhile, at ground level, the constant
grazing gives non-grass species a chance – flowers prosper, and
with them, insects. For all these reasons, grasslands are a
great foraging ground for naturalists.

Downland

Found mainly in Britain and Western Europe, downland is characterized by chalky soil and depends on grazing to avoid being invaded by scrub. Downland is also notable for its abundance of wild flowers – early spring displays of pasque flowers give way in summer to orchids, thistles, and daisies. Insects are abundant, and mammals include rabbits, brown hares, and predatory stoats.

BROWN HARE

PASQUE FLOWER

Grasslands

The wide open spaces of the world's grasslands are dominated by herbaceous plants and grasses, and kept in their natural state by grazing, fire, and long spells of dry weather. Most are home to large herds of herbivores.

Dehesa

A habitat unique to Spain and Portugal, dehesa combines sparse holm and cork oaks and short-sward grass with a rich diversity of flowers, including lavender. It has been created and managed by people and grazing cattle over centuries. Areas left ungrazed revert to scrub and oak saplings take root. Dehesa is home to mammals such as red and roe deer, many reptiles, and birds such as the azure-winged magpie. Large flocks of common cranes feed on fallen acorns during autumn and winter.

AZURE-WINGED MAGPIE

LAVENDER

Alpine grassland

Above the treeline in the high mountains of the Alps, Pyrenees, Carpathians, and in Scandinavia, the ground is covered by snow or subject to hard frosts for much of the year. Generally above 2,000m, the growing season is short in this environment, and alpine grassland has a special mix of plants that can tolerate these conditions. The floristic diversity is reflected in a variety of butterflies, including apollos and many blues and fritillaries. Spanish wild goats, chamois, snow voles, and marmots live here, as do Eurasian dotterel, ptarmigan, ring ouzels, and rock thrushes.

APOLLO BUTTERFLY

ALPINE MARMOT

Steppe

Dry areas of grassland plain, known as steppe, can be found in parts of Mediterranean and eastern Europe, and in Asia. These habitats are known for their contrasting hot summers and harsh winters. After a long winter, the steppes come alive in spring, as crocuses and tulips burst into bloom. Birds such as cranes are found here, along with grazing mammals, including the onager.

DEMOISELLE CRANE

ONAGER

WILD TULIPS

BEE
ORCHID

GREAT GREEN
BUSH CRICKET

CINNABAR
MOTH

Visit downland on a sunny day, when insects become active in the warmth. You will see butterflies feeding on flower nectar.

Downland walk

A dramatic display of flowering plants makes chalky downland the perfect habitat for a wealth of wildlife, with an especially diverse and noticeable array of insects.

In summer, the sky above flower-rich downland is full of flying insects and the sound of birds singing. Small mammals, such as voles and shrews, run through tunnels in the grass, always on the lookout for a predatory stoat or weasel. At ground level, the plants themselves are teeming with life, from tiny ants to bumble bees, while beetles and crickets can be seen perching on stems and leaves.

Keep an eye out for beetles in the grassland turf. Their boring and burrowing activities help keep the grassland ecosystem in good shape.

DARK GREEN
FRITILLARY BUTTERFLY

CENTAURY

Always examine flowers carefully. You may spot a crab spider or ladybird hiding among the petals or in the centre of the flower.

DOR BEETLE

CHALKHILL BLUE BUTTERFLY

Lie down carefully on the grass to see flowers at close quarters and enjoy the scent of herbs such as thyme.

WILD THYME

In the grass

A variety of smaller habitats within grassland all support interesting insect/invertebrate communities, which thrive among the rich diversity of plants.

Life in the rough

Insects live in all layers of the grassland from the turf that is kept cropped by rabbits and other grazers, through the tangled jungle of grasses and herbs, up to the flowering heads of taller plants, where many beetle species can be found. Each species has its own special requirements and many live within a surprisingly small area. Some are fiercely territorial, defending their own patch against intruders, while others are more nomadic, constantly on the move to find food or a mate. When exploring grassy slopes and banks, look closely for ground-foraging beetles and spiders' webs among the grass stems. Also check for snails that thrive on chalky grassland soils and don't forget to look up – the air will be full of butterflies, day-flying moths, hoverflies, and bees.

MALACHITE BEETLE

SINGING INSECTS

Male crickets and grasshoppers use their body parts to "sing". Crickets rub their wings together, while most grasshoppers chafe their hind legs against their forewings. This is known as stridulation and is the characteristic sound heard in grasslands during warm weather. Its purpose is to attract potential mates, who detect the singing via special receptors.

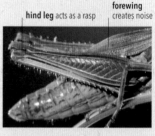

hind leg acts as a rasp

forewing creates noise

GRASSHOPPER

NIGHT LIGHT
Glow-worms can be found worldwide in grassland, and are most visible after dusk. The flightless female attracts males with a light made by a chemical reaction within her body.

SPIDER RELATIVES
Look for harvestmen as they scramble through lower vegetation. They have no silk or venom glands (unlike spiders), so use their long legs to find and trap insects.

BUTTERFLY WATCHING

Although certain butterflies will fly in overcast conditions, most only come out in the sun. The peak time for butterfly activity is mid-morning to mid-afternoon. Choose your site carefully: a sheltered location is best, with plenty of flowering plants to attract the butterflies and warm rocks and grassy banks where they can bask in the sun. Try to get a close look at them as they rest.

ADONIS BLUE BUTTERFLY

BUTTERFLY NET

Remarkable life cycle

Female large blue butterflies lay their eggs on wild thyme flowers. Caterpillars hatch from the eggs and feed on the flower seedheads. When they reach a certain size, they drop to the ground. Red ants, fooled into thinking the caterpillars are their own larvae, take them to their underground nest. Once inside, the caterpillars change their diet and start eating the ant larvae. The following spring, they emerge from the nest burrow as adult butterflies, and the cycle begins again.

USING A QUADRAT

To get a good idea of local plant and invertebrate diversity drop a sampling square, known as a quadrat, randomly on the ground. Note the number of plant and ground-dwelling insect species in each square of the frame.

1. WILD THYME

2. ANTS MOVE A CATERPILLAR TO THEIR NEST

ant larva

3. CATERPILLAR BEGINS TO EAT ANT LARVAE

STRIDULATING GRASSHOPPER

Different species of cricket and grasshopper make distinctive sounds which often follow the same pattern – a series of chirps repeated about every two seconds. Use their song to track down the elusive insects on grass stems and flowerheads.

Grassland close-up

Grasslands provide a rich habitat for plants, insects, and reptiles. The species will vary with location, but study the ground and grasses to see what you can discover.

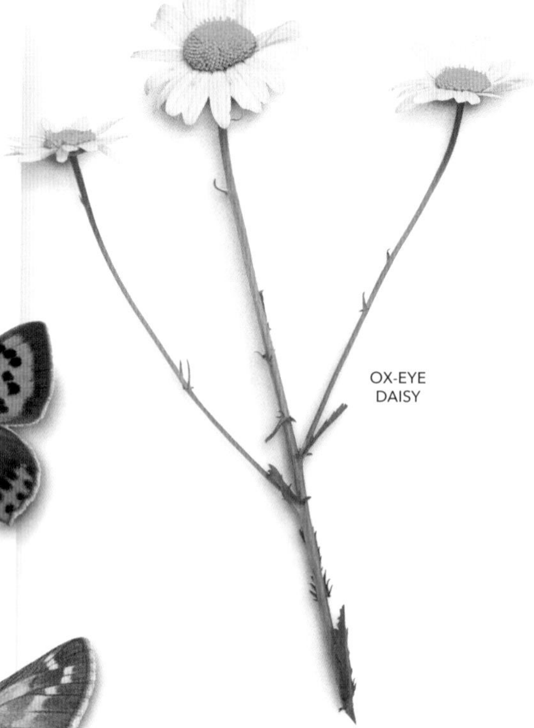

YELLOW SHELL MOTH

Look around flowering plants for butterflies — you may also see their pupae hanging from leaves, twigs, or stems.

SMALL COPPER BUTTERFLY

OX-EYE DAISY

LARGE WHITE BUTTERFLY

SILVER-SPOTTED SKIPPER BUTTERFLY

A closer look through dense grass will reveal that it is also made up of wild flowers, which come into bloom at springtime.

Check grass stems for perching insects and invertebrates, such as grasshoppers.

GREEN TIGER BEETLE

Take care when walking through long grass, as snakes and slow worms may be sheltering there.

HARVESTMAN

TERMITE

GREAT GREEN BUSH CRICKET

SLOW WORM

YELLOW CHAMOMILE

SAW WORT

BURDOCK
SEED HEAD

REST-
HARROW

YELLOW
BEDSTRAW

MUSK
THISTLE

TUFTED HAIR
GRASS

QUAKING
GRASS

COUCH
GRASS

GREEN-
WINGED
ORCHID

Mountain and hillside

Those of us with a hunger for wilderness gravitate to these environments because their slopes often have restricted cultivation, and so they seem less "bruised" by the **hands** of **humans**. But in Europe, at least, our perception of what appears natural is often skewed by idealized landscapes. If we really want to explore these habitats, perhaps we should revel in an exploration of the ecologies and behaviours of the unique assemblage of specialized plants and animals that live in these precipitous places: species for which the edge – and sometimes the void beyond – is a comfortable home.

Plateau

These bleak, windswept places are covered with snow for much of the year and have a similar climate to the Arctic tundra. The plants that thrive here, such as purple saxifrage, grow close to the ground in thick mats and have small leaves to reduce water loss. Plateaus have few predators, making them a haven for animals that can endure the conditions, such as the ptarmigan with its thick, downy plumage.

PURPLE SAXIFRAGE

PTARMIGAN

Mountains

The world's highest places present formidable challenges, with scorching days, fiercely cold nights, and some of the most extreme weather on Earth. The plants and animals that live here need to be superbly adapted to survive.

Rocky pinnacles

Even the bare rock of mountain pinnacles can support life. Freezing and thawing creates cracks in rocks, where you can find plants such as edelweiss, glacier buttercup, and Alpine rock jasmine. The last has even been found growing near the summit of the Matterhorn in the Alps. Sure-footed ibex, chamois, marmots, mountain hares, and ptarmigans can survive at these altitudes, where the ground is covered in snow for much of the year.

EDELWEISS

MOUNTAIN HARE

Forested slopes

Below the treeline, forest cloaks the slopes of many mountains, with coniferous trees dominating, especially pines, firs, spruces, and deciduous larches. The forest can be dense, making it hard to see the animals that live there, but lucky observers may see red deer or even a beech marten or brown bear. The many birds include black grouse, woodpeckers, spotted nutcrackers, and citril finches.

STONE PINE

BEECH MARTEN

Volcano

If you visit an extinct or dormant volcano such as Vesuvius in Italy or Teide on Tenerife, in Spain's Canary Islands, you will notice that the lower slopes are flushed with plant life. That is because the lava and dust that the volcano erupted released many nutrients, creating rich soils. When lava flows from an active volcano, such as Mount Etna on Sicily, all the vegetation in the path of the lava is destroyed. But it does not take long for specialist plants to colonize the area again. And once they are established, lizards and birds move in.

ALPINE CHOUGHS

ITALIAN WALL LIZARD

ALPINE BEES
Bees can nest as high as 2,400m in the Alps. Their furry bodies help to maintain their core temperature.

MARMOTS
Large rodents of the squirrel family, marmots hibernate through the winter in large burrow systems. They have a loud, piercing whistle.

SUMMIT/ NIVAL

GLACIER

ICE FALL

ALPINE

ALPINE MEADOW

SUBALPINE

Living with your head in the clouds

Mountainous areas are home to a range of diverse habitats, with very specialist plants and animals exploiting every niche.

Mountain zones

A vast range of elevations and temperatures means that mountains have numerous microclimates, home to very different plant species. At the base of the mountain, a deciduous woodland gives way to a line of conifers, which thrive in the cold. Above this treeline, the alpine tundra, an area subjected to intense sunlight and wind, is mainly home to small-leaved, low-growing plants. In some sheltered areas of the tundra, wild flowers take root, forming alpine meadows. The tundra finally merges into the scree and rock of the higher slopes, where only lichens grow.

MOUNTAIN PINE
In Europe, the altitude to which trees can successfully grow is marked by a distinctive line of mountain pines.

SUCCULENT GROWTH

"Rain deserts" are common in mountains. Despite heavy rain and snowfall, almost constant wind means that evaporation rates are high. And the high rainfall leaches nutrients out of the poor soil. Plants such as these houseleeks and other succulents counter these problems by storing water and nutrients in their thick, fleshy leaves. They can thrive even on rocky mountain ledges in the Alps with very little soil and loads of sunshine.

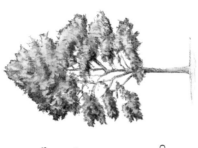

MOUNTAIN CAPRIDS
Nimble and sure-footed, caprids such as chamois live on steep slopes and rugged ground, where they can outrun precators.

MOUNTAIN BUTTERFLIES
Only seen flying for a few weeks in the summer, butterflies, such as the mountain ringlet, overwinter as larvae, buried deep in grass tussocks.

ALPINE NEWTS
Look for these in the grass and around water on mountain heaths in the summer. You won't find any in snowy winters though, as they will be hibernating.

WOODLANDS

MONTANE

Life at high altitude
Life above 3,000m has the added challenge of a lack of oxygen. Mountain birds, such as Alpine choughs, and mammals such as marmots, show various adaptations to get around this problem. The latter have more efficient haemoglobin in their blood, and birds have special sacs that direct air back through the lungs before it is exhaled, to extract as much oxygen as possible. These adaptations allow mountain animals to live normal lives in as little as one third of the oxygen found at sea level.

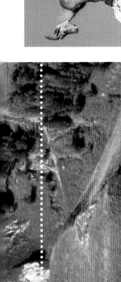

LIVING THE HIGH LIFE
Golden eagles have a high concentration of myoglobin in their muscles, enabling a more efficient uptake of oxygen into the blood.

GETTING AROUND THE MOUNTAIN ZONES
The most important thing to remember about exploring the mountains is to be prepared. The weather is not only more severe, but it can change very suddenly and catch you unaware. Always tell someone where you are going and when you will be back, take a map and compass, and keep abreast of weather forecasts. It is also important to know your limits: set a time by which you need to turn back in order to make it back down safely before dark. Unless you are an experienced mountain climber it is not advisable to attempt hiking in steep or rocky terrain outside of marked paths.

HAVE FUN UP HIGH
Hiking or mountain biking are great ways to get around the mountains, but always remember to stick to designated paths.

Mountain plants

Mountain winters are long and severe, the short summer months bringing an explosion of colour as plants rush to grow and flower while they can.

Life above the treeline

The small alpine plants that grow above the treeline have adapted to low temperatures and humidity, frost and ice, increased winds, and a short growing season. Where there is enough soil, tussock grasses, shrubs, and low trees dominate – larger plants with bigger leaves cannot tolerate the dessication caused by the high winds. You may notice that some plants in this zone have tough, hairy leaves; this is to reduce moisture loss and minimize the effects of frost and ice. Others have special red pigments that can convert the sun's light into heat.

MOUNTAIN LAUREL
This shrub grows on rocky mountain slopes. The leaves are retained year on year and are quite toxic, which prevents them from being eaten by passing animals.

RHODODENDRON

Despite their huge success as garden plants around the globe, the rhododendron is an alpine plant with most species originating in the Himalayas. In their native environments, rhododendron flowers attract scarce mountain pollinators with an array of colours – white, yellow, pink, scarlet, purple, and blue. However, outside of their native range, where they have spread from gardens, rhododendrons often cause problems. They spread quickly and outcompete native plants.

RECURRING BLOOMS
Some meadow flowers, such as the alpine sow-thistle, are perennials, meaning that a new plant grows from the existing root every year.

RED HELLEBORINE

LESSER BUTTERFLY ORCHID

Mountain meadows

Look around you in the mountains and you will see that the various habitats often occur in a patchwork. This is due to prevailing winds, exposure to the sun, hillside location, soil consistency, and underlying rock type, all factors that affect plant growth. Alpine meadows grow in the most favourable habitats, where sediments from weathering rocks create soils capable of sustaining grasses and wildflowers. Most of them store up energy to last them through the harsh winters and flower briefly. Many are dwarfed and stunted by their environment.

ALPINE MEADOW
Alpine plants have a short flowering season. It may take some years to build up enough energy to flower, but when they do it can be a spectacular sight.

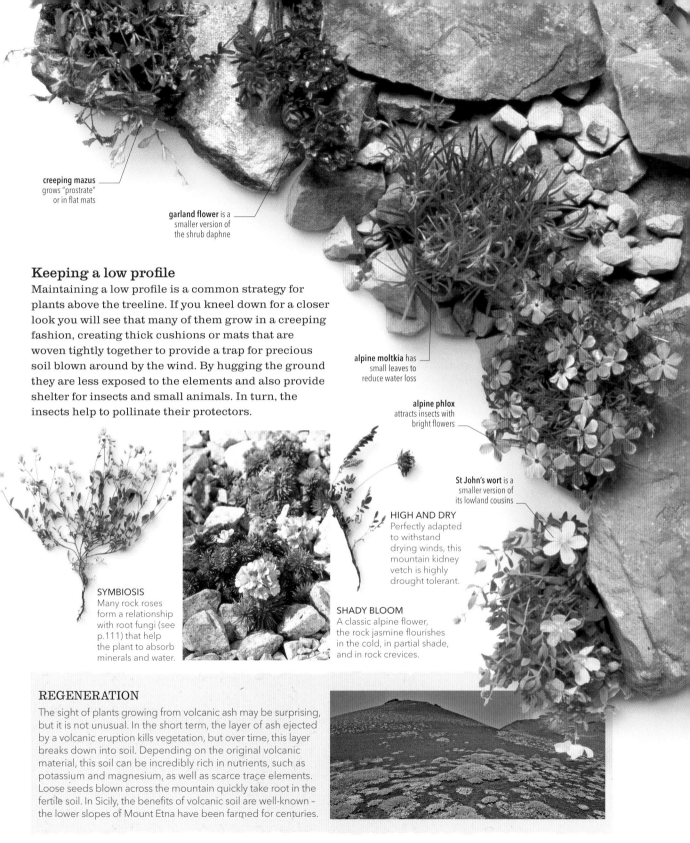

creeping mazus grows "prostrate" or in flat mats

garland flower is a smaller version of the shrub daphne

Keeping a low profile

Maintaining a low profile is a common strategy for plants above the treeline. If you kneel down for a closer look you will see that many of them grow in a creeping fashion, creating thick cushions or mats that are woven tightly together to provide a trap for precious soil blown around by the wind. By hugging the ground they are less exposed to the elements and also provide shelter for insects and small animals. In turn, the insects help to pollinate their protectors.

alpine moltkia has small leaves to reduce water loss

alpine phlox attracts insects with bright flowers

St John's wort is a smaller version of its lowland cousins

HIGH AND DRY
Perfectly adapted to withstand drying winds, this mountain kidney vetch is highly drought tolerant.

SYMBIOSIS
Many rock roses form a relationship with root fungi (see p.111) that help the plant to absorb minerals and water.

SHADY BLOOM
A classic alpine flower, the rock jasmine flourishes in the cold, in partial shade, and in rock crevices.

REGENERATION
The sight of plants growing from volcanic ash may be surprising, but it is not unusual. In the short term, the layer of ash ejected by a volcanic eruption kills vegetation, but over time, this layer breaks down into soil. Depending on the original volcanic material, this soil can be incredibly rich in nutrients, such as potassium and magnesium, as well as scarce trace elements. Loose seeds blown across the mountain quickly take root in the fertile soil. In Sicily, the benefits of volcanic soil are well-known – the lower slopes of Mount Etna have been farmed for centuries.

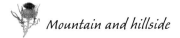
Scaling the heights

Mountain mammals inhabit a precarious environment. However, they are protected by its remoteness.

Survival in the mountains requires athletic sure-footedness and an ability to survive in one of the most extreme habitats in the world. In return, the environment offers protection, a potential lack of competition, and even an escape from parasites and biting insects. For example, in the Alps, chamois and ibex migrate to higher altitudes in the summer months to feed on new plant growth. Mammals that rely on the mountains are highly adapted and thus particularly at risk from the effects of climate change. As temperatures increase, they are forced higher and higher up the slopes and may eventually have nowhere left to go. The best way to spot mountain mammals is with a good pair of binoculars. Find a safe, comfortable spot with a good view and slowly scan the mountainside. Look for any sudden movement or for an ibex or mountain goat balanced on a knife-edged ridge.

MOUNTAIN PREDATORS

In European mountains, most of the mammalian predators live and hunt below the tree line. Eurasian lynxes live up to the tree line in the Alps and Carpathian Mountains and generally prefer to hunt in the dense forest rather than on the higher open slopes. The same is true of wolves. Above the tree line, however, most of the large hunters are birds of prey, including golden eagles and eagle owls, which grab their mammal and bird prey on the ground.

The eagle owl has a wingspan of 1.7m and has powerful claws for seizing prey.

SURE-FOOTED SHIFTERS
Mountain caprids have hooves with sharp rims for lodging in small footholds, and a small rubbery pad between them for improved grip.

Mountain birds of prey

Birds of prey can be found in a variety of habitats, but they are nowhere more at home than in the skies above mountain ranges, soaring on the strong currents of air.

Mountain hunters

Look up while you are hiking in the hills and mountains and you might see a large bird of prey (raptor) soaring high up in the sky. While it may seem effortless, it actually takes a huge amount of energy to get their bulky bodies airborne. Raptors get around this problem by nesting on rocky ledges, and can launch themselves into the air using gravity to give them momentum and then lift. Their keen eyesight means they can scan vast areas of land for prey or carrion – a great advantage on a mountainside where food may be scarce.

THERMALS
Raptors ride thermals in tight upward circles to stay within the current. They may need to flap their wings to move to the next thermal.

Mechanics of flight

The key to soaring flight is rising columns of warm air called thermals. They are created when warm air from the ground rises. Large birds of prey and vultures look for smooth, dark areas, such as ploughed fields or roads, which absorb more heat and thus create more dramatic thermals. In turn, updrafts occur when winds hit the mountains and are forced up. Using both, raptors can soar for hundreds of kilometres, hardly using any energy at all.

FINE FEATHERS

A remarkably versatile body-covering, the feathers of raptors are adapted to a wide range of purposes. Primary flight feathers are the largest, outermost feathers of the wing - they propel the bird forwards then provide lift to keep it aloft. Contour feathers give the bird's body an aerodynamic, streamlined outline, while down feathers provide insulation.

DOWN FEATHER

CONTOUR FEATHER

FEATHER COLLECTION
Look for discarded feathers, they will tell you about the area's birds even when none can be seen.

FEATHER STRUCTURE
Flight feathers have a central, hollow shaft and many side branches called barbs. These hook together to create a solid surface.

each barb has hundreds of barbules

barbules "zip" together

FLIGHT FEATHER

Mountain birds in flight

Egyptian vultures have a wingspan of up to 1.7m

VULTURES
Vultures, such as this Egyptian vulture, have very large wingspans. They wheel in circles on thermals (columns of warm, rising air) looking for carrion.

primary feathers bend up at the tip

EAGLES
Fierce hunters, eagles capture prey by approaching it from behind. Golden eagles are recognizable by the shallow v-shape in which they hold their wings.

FALCONS
Many falcons, including the peregrine (pictured) and the merlin, live in mountainous landscapes. The fastest of all raptors, peregrine falcons can attain speeds of 320kmph when diving.

thin tapered wings enable rapid flight

BEARDED VULTURE
This large bird's diet includes bones that it carries up high, dropping them onto rocks to get to the marrow inside.

Mountain close-up

Mountain wildlife will vary geographically, but you'll also notice a change in plants and animals as you move from one mountain zone to the next.

COULTER PINE NEEDLES

A line of conifers marks the highest point at which trees can grow.

APOLLO BUTTERFLY

Mountain butterflies migrate to higher elevations daily, returning to the lower slopes at night.

COULTER PINE CONE

PINE CONE AND SEEDS

HERMIT BUTTERFLY

SLATE

LICHEN

Mountain insects and spiders seek shelter among low-growing plants.

HORSEFLY

Rocks and fossils provide a glimpse into changes in Earth's geology.

SCAPHITID AMMONITE

COMMON WASP

ORB SPIDER

WHITE GRANITE

RED GRANITE

GRANITE WITH LARGE CRYSTALS OF QUARTZ, MICA, AND FELDSPAR

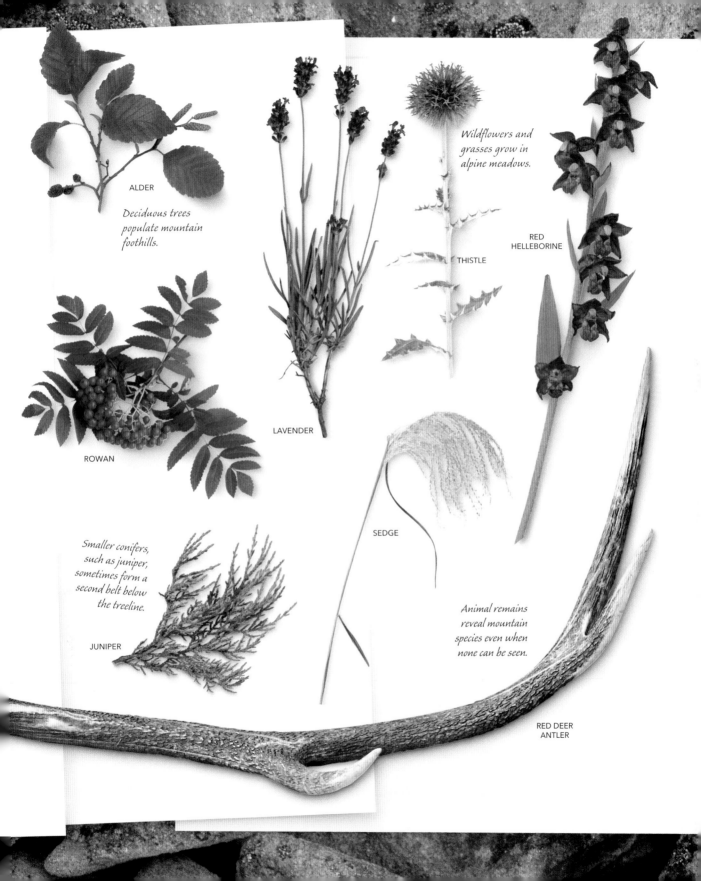

ALDER

Deciduous trees populate mountain foothills.

LAVENDER

ROWAN

THISTLE

Wildflowers and grasses grow in alpine meadows.

RED HELLEBORINE

SEDGE

Smaller conifers, such as juniper, sometimes form a second belt below the treeline.

JUNIPER

Animal remains reveal mountain species even when none can be seen.

RED DEER ANTLER

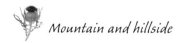
MOUSE-EARED
BAT

Life in the underworld

Caves are particularly prevalent in limestone mountains, formed by chemical action and erosion. The conditions inside are usually fairly constant, and they provide shelter for a multitude of unusual creatures.

Cave dwellers

Animals that live within caves fall into three categories. Troglobites live their whole lives in caves and never come out. Troglophiles, such as the cave salamander, favour caves, but may also live elsewhere. Trogloxenes use caves for shelter or for certain parts of their lifecycle, but also venture out into the light. Classic trogloxenes include a huge number of bat species that roost and hibernate in caves.

CAVE VISITORS
Some animals end up in caves by accident, perhaps swept in by flash floods. Frogs have been found deep underground, apparently thriving!

STALACTITES AND STALAGMITES

Look around a cave and you might notice what look like "stone icicles". These are formed by acidic water, carrying dissolved limestone, dripping through the roof of the cave. Some of the dissolved minerals are left behind and eventually form stalactites, which hang from the roof of the cave and stalagmites, which grow up from the floor where the drips land. This may take tens of thousands of years, and the two may eventually connect as a column.

ANCIENT STALAGMITES AND STALACTITES

SPIDERS
Some cave-dwelling spiders, such as European cave spiders, are thought to be photophobic – averse to the light.

MOTHS
While many moths overwinter as pupae, some, such as the herald (pictured) and tissue moths, hibernate as adults in caves.

CAVE ARTHROPODS
Arachnids such as this pseudoscorpion do well in cave environments. Unlike true scorpions, they do not have a tail with a stinger.

EXPLORING CAVES

Caving is one of the most exciting realms of exploration in nature. Often a tiny entrance will lead to vast cave systems with kilometres of passageways and rooms as big as a sports stadium. However, caves can be very dangerous places – it is easy to lose your footing on loose or uneven rock, and heavy rain can lead to flash floods. Make sure you're fully prepared, and check weather conditions before any expedition into a cave system. If you are new to caving, only attempt it with an experienced guide.

Creatures of the deep

The world's deepest, darkest places are inhabited by some of the strangest-looking animals on the planet. These creatures have adapted to live in the darkness and have evolved to suit their surroundings with useless eyes, extended tactile limbs, antennae for feeling their way around, and an increased sensitivity to air pressure and temperature. As food and oxygen can be scarce underground, troglobites often have low metabolisms and long lifespans. Olms, which live in cave systems in the Dinaric Alps, may live for more than 60 years.

OLM

WINDOW TO THE WORLD
While much of the world underground is barren, cave entrances are a veritable haven for life – sheltered and safe, but with easy access to the outside world.

PLANTS
The cave systems themselves are too dark to support plant life, but cave entrances are often alive with shade-loving plants, known as sciophytes. Sheltered from the wind, they thrive in moist conditions.

OPPOSITE-LEAVED SAXIFRAGE

HERB ROBERT

WOOD SORREL

Lake, river, and stream

There is an almost incomprehensible range of scale in these habitats. Some lakes are sea-sized, and some massive rivers invisibly merge with oceans, yet they also vary throughout their latitude and altitude, as well as in response to the environment beyond their banks. No matter what the location, however, the thirst for a freshwater lifestyle has led to a wonderful richness of species – and for many of us, the humble garden pond forms a perfect doorway to the discovery of this abundance. Lie on your belly and you can peer into the process of metamorphosis, marvel at a web of life linking predators, prey, and plants, and simply enjoy a range of species very alien to yourself.

Upland streams

These turbulent, rocky waterways flow quickly in places, but most have quieter stretches as well. Waterfalls and runs are interspersed with pools, which are home to stonefly and caddisfly larvae, as well as small fish. Birdlife includes wagtails and diving specialists, such as dippers. Few plants can grow in the fast water itself, but ferns and mosses cling to the banks.

WHITE-THROATED DIPPER

Lakes, rivers, and streams

Freshwater habitats are some of the richest in terms of wildlife. The animals and plants that live in them vary, not only according to geography, but also to water chemistry and the speed of water flow.

Lowland rivers

Lowland rivers flow more gently than upland streams, and host a greater range of species. Plant life often grows thickly in the water and on their banks. Mayflies can be seen swarming around the water and laying their eggs on its surface, and mammals, such as beavers, make their home at the waterside in dams made of branches and mud.

MAYFLY

EUROPEAN BEAVER

Lakes

Lowland lakes are full of nutrients and support a variety of animals, including dragonflies, which lay their eggs in and around the water, and bottom-feeding fish, such as carp. In upland lakes the water contains fewer nourishing elements and fewer species of fish. However, diving birds and ducks do make these lakes their home.

MALLARD DUCK

DRAGONFLY

Ponds

Ponds are usually abundant with nutrients, such as nitrogen and phosphorous, and often full of plant life. Many ponds are cut off from streams or rivers, and the animals that live in them are either seasonal visitors or are often introduced. Most insects, such as pond skaters, have wings.

CANADIAN PONDWEED

POND SKATER

Swamps, bogs, and fens

These waterlogged habitats are found in upland and lowland areas. Swamp describes a wetland with continuous water cover. Bogs and fens have peaty soil – bogs are acidic and fens are neutral or alkaline. Vegetation in these habitats includes mosses, sedges, and reeds. Inhabitants range from dragonflies to amphibians, snakes, water birds, and mammals such as water voles.

COOT

EUROPEAN POND TURTLE

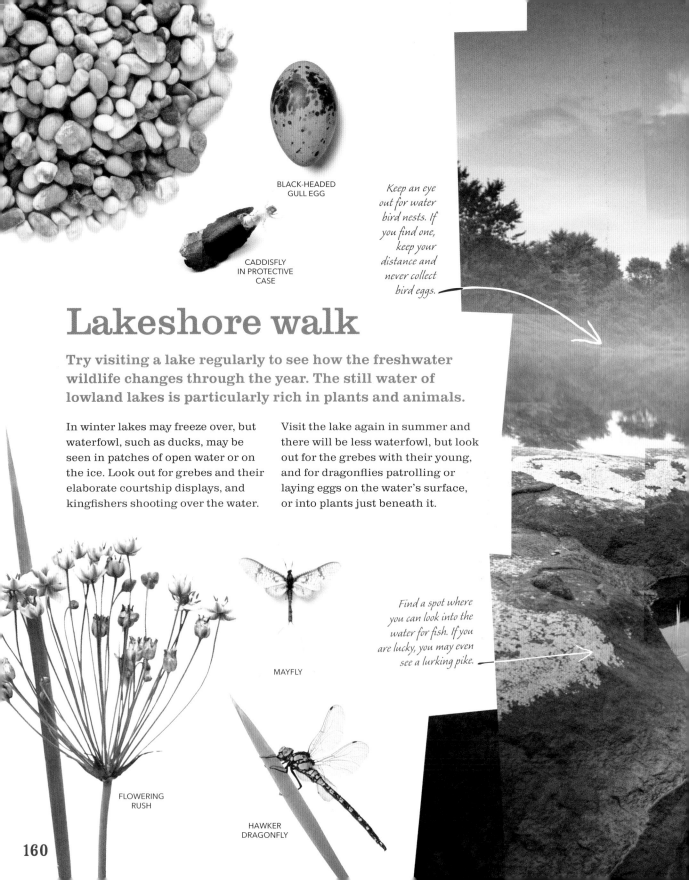

BLACK-HEADED
GULL EGG

CADDISFLY
IN PROTECTIVE
CASE

*Keep an eye
out for water
bird nests. If
you find one,
keep your
distance and
never collect
bird eggs.*

Lakeshore walk

**Try visiting a lake regularly to see how the freshwater
wildlife changes through the year. The still water of
lowland lakes is particularly rich in plants and animals.**

In winter lakes may freeze over, but
waterfowl, such as ducks, may be
seen in patches of open water or on
the ice. Look out for grebes and their
elaborate courtship displays, and
kingfishers shooting over the water.

Visit the lake again in summer and
there will be less waterfowl, but look
out for the grebes with their young,
and for dragonflies patrolling or
laying eggs on the water's surface,
or into plants just beneath it.

*Find a spot where
you can look into the
water for fish. If you
are lucky, you may even
see a lurking pike.*

MAYFLY

FLOWERING
RUSH

HAWKER
DRAGONFLY

EMPEROR
DRAGONFLY

Check the surface
for damselflies and
dragonflies laying eggs.
Mating damselflies
may be seen on
waterside and
emergent vegetation.

BULRUSH, OR
REEDMACE

Take a closer look at
aquatic plants and
you might see a
dragonfly perched
on a stem or leaf.

161

Life of a river

From their small streams to vast coastal estuaries, rivers carve the landscapes through which they flow.

Stream to river

Many rivers are born in higher areas of land, or uplands. Rainwater, melting snow, and water oozing out of bogs trickles into streams. As they flow downhill, these streams meet other streams and a river is formed. Further downstream, a river may join other rivers. Some rivers begin in the lowlands; their water comes from natural springs that rise from subterranean water stores, such as chalk formations. Most rivers make their way to the sea, or into a lake, changing their character – and the animals and plants that depend on them – along the way.

Upland streams and rivers

Nutrients, such as phosphorus and nitrogen, are harder to come by in these bubbling, rocky waters than in the lowlands, so you will see fewer plants and animals here. But keep a lookout for dippers – these short-tailed birds "dip" and will go underwater, searching for food.

WATER WALKER
Dippers are well adapted to upland river life. This plump white-throated dipper is found throughout Europe.

Middle reaches

Here, you should notice calmer water, intermittent rocky stretches, and a greater variety of plant and animal species. Try to spot diving ducks, such as the goosander – a duck with serrated bill edges that help it grip slippery fish. On sandy, muddy, or gravelly sections, look out for the white flowers of water crowfoot, floating on the water's surface. In Europe, you might see beautiful demoiselle damselflies – though only between May and September, during their flight period.

FLYING JEWELS
Demoiselles are exotic-looking damselflies with tinted wings, which live by rivers and streams.

BOTTOM-DWELLER
A sleek, silvery fish with a colourful dorsal fin, the grayling searches riverbeds for larvae and other food.

BRINGER OF LIFE
A fast-running mountain river reveals how water enriches the landscape. Note how trees line the river's rocky banks.

FISH-WATCHING

Fishers wear polarizing glasses to make it easier to see fish in the water. Try wearing a pair to help you spot what's living in your local rivers. Similarly, polarizing filters for camera lenses will help you better record what you see.

POLARIZING SUNGLASSES

Lowland rivers

Nutrient-rich lowland rivers typically support more species than higher stretches, and these are the rivers most of us know best. The type of species that live in and around them is influenced by chemistry – more alkaline waters can be especially rich. Pollution from farming or sewage works can reduce the diversity and number of species a river supports.

SWIMMING SNAKE
Grass snakes are good swimmers. They can reach lengths of 1.2m or more, and feed mainly on toads and frogs.

PATIENT HUNTER
You will often see herons, such as this grey heron, waiting or wading slowly, while looking for fish to eat.

Estuaries

Large rivers flow into the sea at estuaries. At low tide, mudflats are loaded with tiny snails, crustaceans, and other invertebrates that provide rich pickings for shorebirds and wildfowl (see pp.206–07). Estuaries are not always easy (or safe) to explore, so take great care when watching from the edges.

EXPERT FISH-CATCHER
Cormorants are skilled fishers, on estuaries and elsewhere. You can often spy them spreading their wings to dry off after a dive.

LIFE CYCLE OF A SALMON

Atlantic salmon lay their eggs in gravel on stream and riverbeds. The fish that emerge stay in fresh water for one to five years, sometimes more. Then they head downstream to the North Atlantic, where they remain for about one to four years. To spawn, they leave the sea and head back upriver. Most returning salmon find their birth stream, using their sense of smell to help find their way. Most spawn just once in their life, but a minority repeat their journey, spawning up to four times in total.

INCREDIBLE JOURNEY
On their way to spawn, salmon jump weirs and waterfalls, and clear heights of over 3m.

Riverbank

For the best riverbank experience, try a quiet walk, or just sit and watch – preferably when there aren't any other people around.

Wet and woody

The saturated soil of a riverbank supports an abundance of plant life, which in turn provides protected areas for insects and invertebrates to reproduce. Mammals, such as otters, also make the

WILLOW
There are many willow species, and some trees are hybrids. Telling them apart isn't easy.

waterside their home, living in well-hidden "holts" within the dense vegetation. Willow and alder trees can be seen along riverbanks. Alders are the only broadleaved trees with cones, and their seeds provide food for many birds. Weeping willows, with their drooping branches, are easy to spot by European rivers.

ALDER
Long male catkins (flower clusters) hang from alders in winter, while shorter cone-shaped female catkins can be seen in spring.

Riverbank-dwellers

Rivers mean fish, so predators, such as otters and minks, thrive here. These mammals all have various adaptations for life at the riverside, such as webbed feet and whiskers, to help navigation through murky waters. You may see an elusive otter – but don't confuse it with the smaller mink. Seeing a beaver is very rare, but you could spot their lodges or dams, made from nearby trees.

SKILLED BUILDER
Beavers are expert tree-fellers: just one family can cut down several hundred trees in a single winter for dams, lodges, and food.

Riverside fishers

Fish-eating birds have two main methods of hunting – stalking and diving. You can see herons stand patiently, or stalk, when hunting. Fish are their main prey, but they will also eat amphibians, reptiles, and insects. Kingfishers perch, watch, then dive in to grab a small fish.

QUIET HUNTER
Heron species vary in size, but all use the "stalk-and-stab or -grab" approach when feeding.

fish swallowed head-first

ragged crest

RAPID DIVERS
You may just see a flash of bright blue as a common kingfisher dives at speed to catch a fish. Back on its perch, the bird will swallow it head first.

1 ADULT BREAKS FREE
The nymph hauls itself out of the water and the new adult breaks out of its final larval exoskeleton.

2 FLUID ENTERS WINGS
Body fluid pumps into the new adult's wings by contractions, to give them their full form.

3 FULL-SIZED ADULT
A damselfly may take a week and a half or longer to become full-sized and ready to mate.

LIFE OF A DAMSELFLY

Damselflies and dragonflies spend most of their lives underwater. Eggs are laid by adults and develop into aquatic nymphs, or larvae. These are fierce predators. To grow, they must shed their skin and may do this more than ten times before they emerge. Some spend five years underwater. Adults may live less than two weeks, but some survive for two months.

Inhabitants

WATER VOLE
Water voles are small rodents found in Europe and Asia. They feed on plants and grasses along the water's edge and also use this material to line their burrows.

DANUBE CRESTED NEWT
This newt lives in rivers, ponds, and lakes in central and southeast Europe. It can reach 13cm (5in) in length, sometimes more.

NIGHT HERON
This black-crowned night heron is the world's most common heron species. It feeds mainly between dusk and dawn on a varied diet, ranging from fish and reptiles to bats.

WATER SHREW
The water shrew can be found throughout much of Europe. It swims very well and, unfortunately for its aquatic-invertebrate prey, has poisonous saliva.

WEBBED PREDATOR
All freshwater otters eat fish as their main food, but some will take birds, small mammals, or frogs. The webbing between their toes helps make them superb swimmers.

Water birds

Swans, geese, ducks, and grebes are just some of the fascinating birds you may see on larger areas of open fresh water.

Waterfowl and other birds

Wetlands attract all types of bird life; many come to feed, or to nest and raise their young within the dense vegetation. Some of the most common birds are known as "waterfowl", a group that includes swans, geese, and ducks. However, this is not the only group of birds that live in this habitat, others include storks and herons. Diversity is the name of the game here – some species build floating nests, others nest in tree holes. Some eat fish, while others feed on invertebrates or plants.

AGGRESSIVE SHOW
Birds like the mute swan make threat displays, including wing flapping and "busking" – where a swimming birds pulls back its neck and lifts its wings.

TAKING OFF
Some water birds, such as this trumpeter swan, need plenty of space for take-off. Watch them run over the surface to help get the required momentum for flight.

long, slender neck

UPENDING
Many water bird species have long necks to allow them to reach underwater plants far down, especially when they upend.

SWAN FOOD
Dabbling birds feed by skimming in shallow water and sieving food and water through filters in their bills.

RED-BREASTED GOOSE
Thousands of red-breasted geese spend the winter on wetlands in eastern Europe, migrating north in spring to breed on Arctic tundra.

GREYLAG GOOSE
Most farmyard geese are descended from greylags. Different subspecies have different bill colours.

Geese

They may have webbed feet like ducks and swans, but geese are adapted to eat plants on land. Their bills suit their tough vegetarian diet, and they walk well because their legs are central to their bodies. Like swans, male and female birds look alike. Usually, they mate for life, breed in the far north, migrate in family units, and winter further south. In some areas, however, Canada geese and the UK's feral greylags can be seen all year round. You will see geese flying in the "V" formation, which is also characteristic of ducks.

WHITE-FRONTED GOOSE
Like all true geese, white-fronts are found only in the northern hemisphere. They breed at high latitudes, but wintering birds can be seen in Europe and the USA.

white patch at base of bill

orange legs and feet

Ducks

Ducks can be divided into divers and dabblers. The mallard is a dabbler, the tufted duck and scaup are divers. Diving ducks propel themselves underwater to feed. To help them swim, their legs are positioned towards the rear, which makes walking awkward. You are less likely to see divers on shallow water than dabblers, as the latter typically skim for food at the surface of shallow waters, or a little lower when they upend. Dabblers' legs are positioned further forward, so walking is easier. Male ducks in decent plumage are usually fairly easy to identify and distinguish from females. Be warned, however – when males moult their flight feathers they adopt a more cryptic plumage, and may look similar to the females.

orange chest-band

COMMON SHELDUCK
This large, boldly marked bird is easy to see in parts of Europe. Males are larger, brighter, and have a red bulge on their bill.

MANDARIN DUCK
This bird nests in tree holes in parts of southern England. This male is much more colourful than the greyish females.

RAISING YOUNG
As is typical among dabbling ducks, this female mallard has sole responsibility for looking after her ducklings. The male probably left when she was brooding her eggs.

Grebes

Grebes are striking birds. Males often resemble females, although some species look very different when they are not in breeding plumage. A grebe's feet are set at its back end, making them superb swimmers when hunting fish, but vulnerable on land. They have lobed toes rather than webbed feet. Floating nests are common, and adults transport their chicks on their backs. It is unusual to see a grebe fly, although most of them can.

chick rides on its parent's back

GREAT CRESTED GREBE
This European grebe is now fairly common. In the 19th century, its feathers adorned women's clothing and numbers plummeted. Flooded gravel pits provided new habitats and helped its UK population recover.

RED-NECKED GREBE
This European species dives under water to catch small fish with its strong, sharp bill.

GREBE COURTSHIP

Finding a partner is something grebes do with style, putting their attractive head ornamentations to good use. These European great crested grebes are performing the "weed ceremony". Partners ascend from the water face to face, then swing their weed-laden bills from side to side. Courtship begins in winter, and also includes head-shaking, which is fairly easy to observe. The North American western grebe is famous for its "rushing ceremony", where birds rush side-by-side and upright over the water's surface.

On the surface

Look carefully at the water in freshwater habitats and you will see that the surface can be alive with a variety of creatures.

Animal adaptations

The water surface is an unusual and fascinating micro-habitat that is inhabited by a variety of specially adapted animals. Pond skaters feed on animals that have fallen in and become trapped. They live up to their name by walking quickly across the surface film; their long legs distribute their weight over the water. Water snails also move across the surface film, but they cling to it from beneath the water. Beetles and water boatmen use the water surface as a temporary filling station, taking on air before and after diving into the water.

SURFACE INHABITANTS
Some animals, such as this water boatman, have special adaptations like long legs and sensitive water-repelling hairs to help them move on water.

SEMI-AQUATIC SPIDER
The European raft spider has water-resistant hairs on its legs to enable it to detect vibrations and run over the water surface after its prey.

TRANSPARENT FLEA
There are many different species of water flea, which are crustaceans. Some use their branched antennae as oars to swim around under the water surface, filtering microscopic, organic food particles out of the water.

SENSITIVE HAIRS
Water-repelling (hydrophobic) hairs stop pond skaters from getting wet and sinking, making walking on water seem effortless. They are common and easy to spot – look carefully for the tiny depressions its feet make on the water surface.

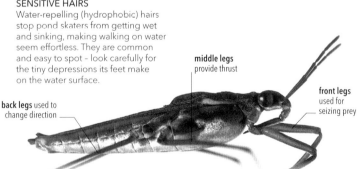

middle legs provide thrust
back legs used to change direction
front legs used for seizing prey

TESTING SURFACE TENSION

The water surface is like a very thin, transparent film that is strong enough to support a small amount of weight before it gives way – this is called surface tension. Watch soap bubbles and you will see clearly the surface layer of a liquid. You can test surface tension with a simple experiment. Put some water in a bowl, take a small sewing needle, and try to float it on the surface. If you are having problems, try floating a small piece of tissue on the water and putting the needle on top of it – it might be easier if a small part of the needle projects over the edge of the tissue. Push the tissue down and away from the needle and, with practice, the needle should float, supported by surface tension. Animals, such as pond skaters, raft spiders, and water measurers are very light and their long, hairy legs allow them to use surface tension to walk on water.

Surface predators

Insects that don't normally live on water, such as flies, can become caught on the surface if they fall in. This is good news for pond skaters, whirligig beetles, water crickets, and raft spiders that all hunt their prey at surface level, and can move across the water at quite a speed to catch it. Great water boatmen also hunt at the water surface, but attack their prey from beneath the water.

NASTY BITE
Great water boatmen should not be handled as they can bite. These predators sense vibrations on the water's surface and attack fish and tadpoles with their penetrating "beak".

OPPORTUNIST
Water measurers catch water fleas that live beneath the water and also eat insects that are trapped on the surface.

PREDATORY FLY
Some brightly coloured, long-legged flies live on the surface film and feed on mosquitoes and other small insects.

SPEED SKATING
Whirligig beetles whizz in circles on water. Their eyes are divided to see predators and prey above and below the water line.

WATER HUNTER
Look out for water crickets' orange markings and watch these insects hunt. When chasing food, they can move at great speed.

Pond dipping

A healthy pond will be teeming with life; pond dipping is a simple and rewarding way to get a closer look at nature.

Viewing pond life

Find a safe, stable spot on the edge and always supervise children closely. Try to use a long-handled net with a fine mesh. To make sure you catch a variety of specimens, don't just dip it into open water, as some creatures live around plants in mud at the bottom.

VIEWING JAR
Put some pond water in a clear glass jar or bucket and empty the contents of your net into it. Let it settle and see what you have caught. Always return your catch to the pond.

caddis flies come out at dusk or night and are often mistaken for moths

frogs need to lift their head above water in order to breathe

webbed hind foot for swimming

damselfly nymph

oar-like hind legs

great water boatmen should not be handled as they can bite – their usual prey include tiny fish and tadpoles

freshwater snails graze on underwater plants, but they need oxygen and return to the surface to take gulps of air

water scorpions have large front legs for catching prey, such as small fish and tadpoles

caddis case of plant material

caddis fly larva surrounded by a protective case that is built with sand and stones

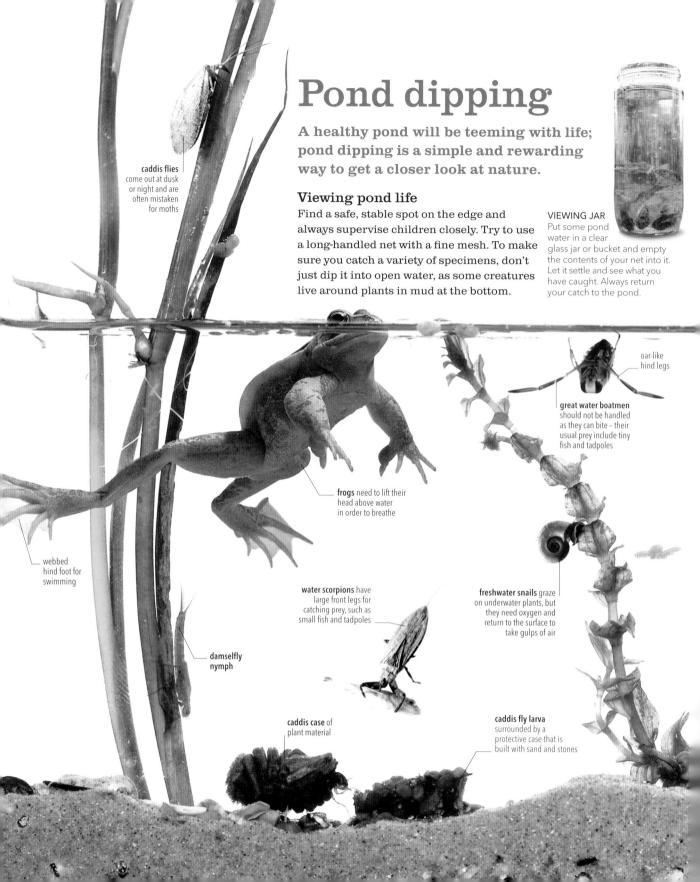

A COURTSHIP DANCE
Smooth newts live in European wetlands and ponds. During the breeding season the female is lured in by the male's spots and vibrating tail.

Breeding season

If you visit a pond during the breeding season of April and May, and shine your torch into the shallow water at night, you might be lucky enough to see courting newts. Pond newts do not spend all their life in a pond, but do return to the water to breed. You may be able to pick out a courting male, as he is typically more "crested" and colourful than the female. He uses his tail to waft pheromones (chemical signals) through the water to the female, and then drops a spermatophore (packet of sperm) near to her.

TADPOLE DEVELOPMENT

The transformation from frog spawn to froglet is fascinating to watch and is an activity you can encourage. The shallow, unpolluted water of a garden pond is ideal to raise tadpoles. Keep the pond well vegetated to provide food and places to hide for the froglets.

1 TADPOLE JUST HATCHED
Tadpoles emerge from the spawn after 30–40 days. They begin by eating the spawn then algae. You can also give them chopped lettuce in icecubes.

2 TADPOLE WITH HIND LEGS
Between six and nine weeks, the tadpole grows hind legs. Provide plants, such as lilies, and branches and rocks as hiding places.

3 TADPOLE WITH ALL LEGS
By about week 11 the front legs are fully developed. Provide plenty of rocks, or a sloped piece of wood, so that they are able to climb out of the water.

4 FROGLET
By week 12 metamorphosis is complete. The froglets will now eat small invertebrates. It could be two or three years before the frog breeds.

fringes on legs propel beetle through water

strong, grasping front legs

great diving beetles are fierce predators that can be up to 3.5cm long – their diet includes newts, frogs, and fish

dragonfly eggs

male sticklebacks are aggressively territorial and build nests they attract females to for egg laying

throat and belly turn red in spring

dragonfly nymph

Swamps, bogs, and fens

If you want to encounter a variety of species in a truly wild setting, try spending some time in one of these fascinating wetland habitats.

Insect-eating plants

The best places to see sundews are waterlogged habitats, such as bogs, which are nutrient-poor as water cannot flow to, or from, them easily. These remarkable plants get the nutrients they need by catching insects. The "hairs" on a sundew's leaf have a gland on each end that secretes a sticky substance. Once an insect is caught, the edges of the leaf gradually curl over, and enzymes help break the insect down. Sundews aren't the only carnivorous plants – bladderworts, butterworts, and waterwheel plants also capture invertebrates in European bogs.

WATER WORLD
You will find plenty of plants even in the soggiest swampland. Reed swamps are dominated by reeds and bulrushes, while cypress trees flourish in swamp forests.

DEADLY VARIETY
There are about 150 sundew species. The great sundew has a large range, and grows in habitats in Europe, North America, and Japan.

fairly long, reddish bill

SMOOTH MOVER
Watch the edge of a European reedbed and you might see a slim water rail. Listen carefully for its shrieks, squeals, and grunts.

Reedbed birds

Some birds have adapted to life in reedbeds – wetlands dominated by reeds. Many rails, such as the water rail, are surprisingly slim, enabling them to get through dense vegetation, and the plumage of bitterns provides great camouflage. Reed warblers weave their nests around reed stems, and as the plants grow, the nest goes up with them.

male shows black cheek stripes

REED RESIDENT
Look for colourful bearded tits in European and Asian reedbeds, where they eat reed seeds during the winter. Their *ping ping* calls might help you find them.

SILENT HUNTER
Bitterns, egrets, and herons, such as this purple heron, breed in marshes. They move quietly around the water's edge, preying on amphibians, fish, and insects.

PRODUCTS FOR PEOPLE

While reeds are still used for thatching, they aren't the only wetland plant to have been exploited by humans. Sedge is also used in thatching, and in some areas, peat is still used as a fuel. Watercress is a bog plant that has been grown commercially in the UK for more than 200 years. It is also cultivated in many other countries, including Italy, the Netherlands, and Spain.

Wetland inhabitants

Swamps, bogs, and fens provide a home for many plants and animals. Those shown here are just a few of those you might encounter on a wetland visit. You won't find them all at every site, of course, but if you go with open eyes and an enquiring mind you should see something special.

1 Bog asphodel flowers in acidic European bogs between July and September. It is a short plant that produces orange fruits.

2 The common reed is a grass found in Europe, the USA, and Asia. It can grow to heights of 3m or more. In bloom, it reveals purple flowers.

3 The cranberry is found in Europe, Asia, and America and has pink flowers, which can be seen between June and August.

4 The yellow or flag iris is a distinctive plant that grows in Europe, the USA, northwest Africa, and parts of Asia. It flowers between June and August and can be over 1m tall.

5 The European common lizard is found in a wide variety of habitats. Look for them soaking up the sun on warm summer days.

6 Grass snakes grow to 1.5m and are particularly fond of wetlands. On hot days, they are often seen swimming close to the surface in freshwater pools.

7 The marsh frog is found in Europe and Asia. This large amphibian can grow to 15cm, and sometimes more. Look for frogs on lily pads.

8 Scarlet darter dragonflies live near the still, shallow waters of swamps in southern Europe. They are about 4cm long.

9 The snipe is a wading bird that can open the tip of its long bill independently to the rest of it. The sensitive tip is used to feel for food.

BOGBEAN
This attractive plant is found in Europe, Asia, and North America. Its pink-and-white flowers bloom between April and June.

173

Coast

The interfaces between land and sea are among the richest habitats on the planet at any latitude, because the mix of the terrestrial and marine generates opportunities for an immense diversity of life. Combine this with a multitude of geological and geographical variables, as well as the resulting range of coastal types, and that diversity expands even further. From mudflats to mangroves, sand to shingle beaches, towering cliff-sides to tidal pools, there is a fabulous array of species living on the edge of land and sea. It's also the place where you can safely explore part of the marine environment without getting too wet!

Beach **p.176**

Cliff **p.194**

Coastal wetland **p.202**

Ocean **p.208**

Sand

Sand forms a range of habitat types. Specialized creatures inhabit tide-washed sand, drawing birds such as sanderlings in to feed on them, while dry sand can blow into vast dunes which, when colonized and anchored by plants such as sea holly or marram grass, form a rich but environmentally challenging habitat for wildlife.

SEA HOLLY

SANDERLING

Beaches

At the front line between land and water, beaches are created and shaped by the actions of the sea, which erodes and deposits sediment. The nature of a beach, however, depends on the geography and geology of the land nearby.

Rock

Waves are a potent force of erosion, wearing away even the hardest of rocks where they meet the sea. On one hand this creates cliffs, but the other result – a rocky beach at sea level – is equally dramatic and the source of endless exploration and inspiration for any naturalist. Each rock pool is a miniature ocean, home to a variety of accessible marine life such as mussels, anemones, and seaweeds.

MUSSELS

SEA-ANEMONE

WRACK

Coral

The gleaming white sand of many tropical beaches and some in the Mediterranean formed from the broken and bleached remnants of coral from shallow-water reefs. At higher latitudes, similar beaches are built from quartz sand and crushed seashells. Both are local beach types that reflect local sources of sediment, as revealed by the shells and bits of coral that wash up on them. All are rich in birdlife. Some, such as little terns, feed offshore and nest on the beach.

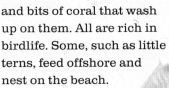
LITTLE TERN

CORAL REEF

MUREX SHELL

Shingle

Shingle is also a product of erosion. Pebbles are deposited as a fringing beach along an exposed coastline, sometimes thrown into ridges by storm waves. While the seaward zone supports annual plants, whose life cycle takes place in summer, more stable shingle supports drought-tolerant perennials such as sea kale and stonecrop. Ground-nesting birds, such as terns, thrive within the mosaic of pebbles and plants.

SEA KALE

STONECROP

SEA BUCKTHORN

Find a patch of sea buckthorn in autumn and you may see birds such as thrushes gorging on the energy-rich fruits to sustain them on their migration.

BUMBLEBEE

Pollinating bees may be scarce in such exposed habitats, so large flowers are needed to attract them. Watch and wait to see what arrives.

Beach walk

Bleak and windswept, a beach in winter resonates with the cries of seabirds. And summer brings a whole new set of sights, scents, and sounds.

Whether made up of sand or shingle, all beaches drain freely, and plants here must be able to cope with drought. Gently squeeze and stroke the leaves and you'll find coatings of wax or hairs and succulent, fleshy stems – all part of the plants' adaptations to preserve water. Large, showy flowers attract pollinating insects, which lure in dragonflies and other predators from their freshwater breeding sites.

YELLOW HORNED POPPY

SOUTHERN HAWKER DRAGONFLY

VIPER'S BUGLOSS

Check out nectar-rich flowers for hawk moths and hoverflies. They provide a vital meal for these migrating insects.

RED VALERIAN

SMALL WHITE BUTTERFLY CATERPILLARS

Tread carefully! Ground-nesting birds such as plovers lay their eggs directly on the beach, relying on camouflage for protection.

PLOVER EGG

SOW THISTLE

Delve into foliage, or search by torchlight at night, to find insect larvae. Susceptible to drought and predation, they hide deep within the undergrowth.

Turning tides

A knowledge of tides is crucial to the exploration of coastal habitats, both for safety reasons and because tides affect everything that lives within them or nearby.

Understanding tides

As the Earth spins on its axis, each point on the planet's surface passes through two high tides once every 24 hours. These bulges of water are formed by the gravitational pull between the Moon and the Earth. High tides occur twice each day at intervals that are a little over 12 hours apart, due to the Moon's orbit changing its position relative to the Earth.

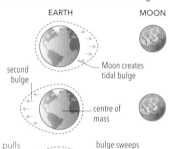

EARTH   MOON

second bulge — Moon creates tidal bulge

centre of mass

bulge sweeps over surface

combined forces

MOON'S PULL
Once the Moon's gravity pulls a bulge of water towards it, a counterbalancing bulge forms on Earth's opposite side as the planet rotates.

READING A TIDE TABLE

Tide tables tell you when high and low tides occur on any particular day, and they normally give the predicted heights of those tides. Learn to use them – they could save your life. But remember that these are only predictions; both the timing and height of tides can be affected dramatically by weather conditions and atmospheric pressure, so remember to keep checking tides visually, too.

TIDE TABLES

4

3

4 Spray zone

This area beyond the tide's reach is still strongly influenced by the sea. Wind-driven salt spray means that the animals here, such as sea slaters, must tolerate a salty environment. On exposed coasts, the spray zone may extend hundreds of metres inland.

SEA SLATER

PERIWINKLE

3 Upper tidal zone

Survival in the upper tidal zone means an ability to tolerate exposure to air and varying salt levels. In hot weather, sea water evaporates and becomes saltier, but in wet weather it is diluted. A few species, including channelled wrack, can tolerate these extremes, others such as anemones retreat into pools.

CHANNELLED WRACK

BEADLET ANEMONE

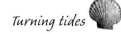

Tidal cycle

In most places, the high to low tide cycle takes a little over 12 hours, but there are also longer-term tidal cycles. At full and new Moon, high tides are higher and low tides lower ("spring tides"), with "neap tides" – lower high and higher low tides – in between. And even longer cycles occur: the highest tides of all occur around the spring and autumn equinoxes.

STRANDED JELLYFISH
When a high tide recedes, it may leave behind some marine animals, like jellyfish.

Take a close look, but don't touch – it can still sting!

HIGH AND LOW TIDE
The constant ebb and flow of tides changes the land- and seascapes of our coastlines. Animals and plants in tidal zones must move or adapt to ever-changing conditions.

POWER OF THE TIDE

The shape and orientation of certain tidal inlets sometimes produces a powerful bore or tidal wave on the rising tide, which may be several metres high and push far upstream. Although a potential pollution-free, renewable energy source, harnessing this power can seriously harm wildlife dependent upon the natural tidal progression.

SURFERS ON A TIDAL BORE

2

1

2 Middle tidal zone

In addition to constant wave action, the middle tidal reaches are subject to alternate dousing with sea water and exposure to air. Brown seaweeds dominate this zone on rocky shores, and you will find animals such as crabs, whelks, and barnacles here.

DOG-WHELK

BARNACLES

EDIBLE CRAB

BROWN SEAWEED

1 Lower tidal zone

Here, sea water dominates and wildlife needs to cope with only short periods exposed to the air. However, wave action is constant, so any seaweed is tough and leathery. Animals such as lugworms survive in burrows, and sea stars anchor themselves with tube feet.

SEA STAR

LUGWORMS AND CAST

Sand and shingle

Coastlines are shaped by the sea. Tides, waves, and currents erode, transport, and deposit material, sculpting diverse landscapes.

Cliffs are eroded by the action of the waves. Rocks are then worn down into ever smaller fragments, which can be picked up by currents and transported along the shore until the strength of the current is no longer able to support them. In this way, transported sediments are sorted into different sizes – the smaller the particle, the further it is carried from where it eroded. Each sediment type supports distinct habitats, and each has a characteristic range of plants and animals.

LONGSHORE DRIFT
You can work out in which direction sediment is being transported by the sea by looking for a build-up of material on one side of a spit of land or barrier.

Examine boulders on the upper shore for barnacles, taking care to avoid unstable rock surfaces.

pebbles are deposited in areas of high current strength, and can be thrown up into ridges by storm waves

coarse shingle is highly abrasive when carried by sea – shingle foreshores are often devoid of life

64–16mm diameter

16–4mm diameter

BOULDERS
Angular chunks of rock that fall from a cliff gradually become rounded, worn down by the sea's continued erosive action as happened to these boulders on the south coast of England.

SEDIMENT
Sorted into size classes by coastal currents, different sediments are deposited in different environments. The finest particles of all, called silt, are laid down only in the most sheltered conditions, such as in the lee of an offshore barrier or in the heart of an estuary, forming mudflats and salt marshes.

ROCK FORMATIONS

Erosion not only produces sediment, but also creates cliffs and other distinctive geological features. Cliffs are rarely uniform – the rocks that are exposed vary in composition. Some are more vulnerable to erosion than others, and different erosion rates can produce large landscape features such as headlands of harder rock. On a local scale, a small weakness in the rock may be attacked by waves, forming a cave and eventually extending right through a headland in the form of an arch. Once gravity takes over, the roof collapses to leave a stack.

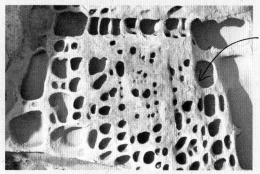

Look for pockets of erosion where the rock was softer.

HONEYCOMB ROCK
Sedimentary rocks vary in hardness as a result of their formation. Water or wind-blown sand often scours the weak points, creating remarkable, almost artistic erosion features.

fine shingle particles hold water and organic matter, which allows invertebrates to move in

sand is foundation of some of our best-loved habitats – dunes for wildlife and beaches for our enjoyment

SPIT
Sand and shingle extends as a spit across the mouth of an estuary in southwest England.

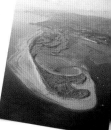

4–2mm diameter

2–0.125mm diameter

DUNES
Carried first by water and then blown by the wind, sand can pile up and form large mobile dune systems, such as the Dune du Pilat in France.

BLACK SAND
Sand's colour reflects the colour of its parent rock. The black sand found on Fuerteventura, in the Canary Islands, formed from volcanic lava.

Rockpooling

Rock pools are a window into a rich underwater world of marine life, otherwise visible only to the diver or snorkeler.

Types of rock pool

Rock pools of all shapes and sizes are revealed when the tide recedes. Deep pools, with overhangs and crevices, have more niches for different plants and animals, and provide shelter from the action of the waves. Shallower pools are easier to investigate, but contain a more restricted range of life.

ROCKPOOL DEPTHS
Shallow pools have greater fluctuations in salinity (salt level) and temperature, making it much harder for wildlife to adapt and survive there.

brown seaweed

sea urchin

DEEP ROCK POOL

SHALLOW ROCK POOL

Anything you lift out of a rock pool must be put back.

STAYING SAFE
Take care on the shore. The rocks can be slippery, the pools deep, sometimes hidden by seaweed – and always keep an eye on the tide.

Life in a rock pool

Every pool is a marine microcosm, home to plants and animals, both predators and prey. The mini-dramas of everyday life and death play out before your eyes, as you peer into the pools at low tide. But with predators around, it pays to be well hidden – you will find much more by exploring among the seaweed fronds and holdfasts, under the boulders, and deep into crevices. Watch rocks closely as some animals, such as crabs, are masters of camouflage.

MAKING A VIEWER

Light reflecting off the surface of water and ripples caused by the wind can make it difficult to see what lies beneath. Sunglasses with polarizing lenses help to reduce the glare, but better still, a simple underwater viewer will reveal all – especially in bright conditions.

1 Take a plastic ice-cream tub and cut out the bottom with a sharp knife.

2 Carefully cover over any sharp edges with strips of waterproof tape, such as duct tape.

3 Tightly roll clingfilm over the bottom of the tub and fasten it in place with an elastic band.

4 Put the covered end of the viewer just under the water's surface, and peer through the box.

Survival strategies

Pools of water at low tide allow marine animals to survive higher up the shore than would otherwise be the case, by protecting them from drying out in the air. But animals on the menu of predators such as starfish must take other defensive measures, including living at the edge of the pool where these predators cannot reach, at least until the tide comes in.

tough ridged surface

twin shells close tightly

SHUT TIGHT
Out of the water, mussels close their shells for protection. However, some shorebirds, such as oystercatchers, have strong bills to prise open bivalve shells.

DRAWING IN
The stinging tentacles of anemones (green, above) open only under water. At low tide, they are withdrawn into the body, leaving a round jelly-like blob.

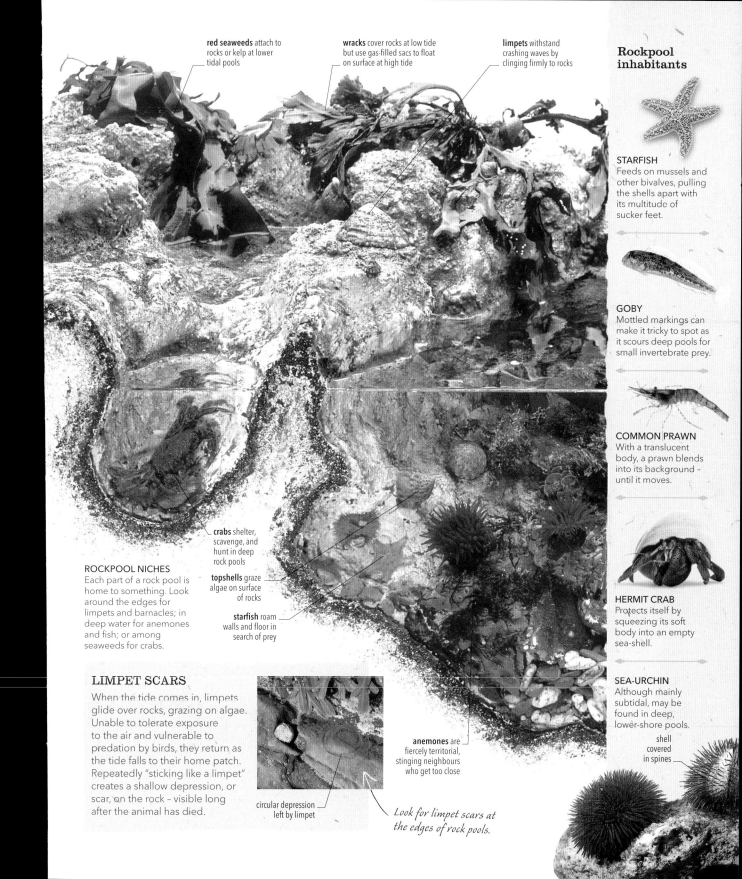

red seaweeds attach to rocks or kelp at lower tidal pools

wracks cover rocks at low tide but use gas-filled sacs to float on surface at high tide

limpets withstand crashing waves by clinging firmly to rocks

Rockpool inhabitants

STARFISH
Feeds on mussels and other bivalves, pulling the shells apart with its multitude of sucker feet.

GOBY
Mottled markings can make it tricky to spot as it scours deep pools for small invertebrate prey.

COMMON PRAWN
With a translucent body, a prawn blends into its background – until it moves.

HERMIT CRAB
Protects itself by squeezing its soft body into an empty sea-shell.

SEA-URCHIN
Although mainly subtidal, may be found in deep, lower-shore pools.

shell covered in spines

crabs shelter, scavenge, and hunt in deep rock pools

topshells graze algae on surface of rocks

starfish roam walls and floor in search of prey

ROCKPOOL NICHES
Each part of a rock pool is home to something. Look around the edges for limpets and barnacles; in deep water for anemones and fish; or among seaweeds for crabs.

LIMPET SCARS
When the tide comes in, limpets glide over rocks, grazing on algae. Unable to tolerate exposure to the air and vulnerable to predation by birds, they return as the tide falls to their home patch. Repeatedly "sticking like a limpet" creates a shallow depression, or scar, on the rock – visible long after the animal has died.

circular depression left by limpet

anemones are fiercely territorial, stinging neighbours who get too close

Look for limpet scars at the edges of rock pools.

Shorebirds

Wherever you are in the world, some of the greatest concentrations of bird life can be found along shorelines, attracted by abundant food.

Shoreline specialists

Birds gather wherever there is food. The twice-daily tides that wash our shores bring in nutrients that support a rich and diverse food chain. At the top of this food chain are birds. As the tide retreats, waders and gulls throng the shoreline, probing beneath the surface for invertebrates. Many waterbirds breed around the coast – sometimes in vast colonies – while others make use of shorelines as part of their annual migrations.

STIFF COMPETITION
To reduce competition for food resources, different bird species vary in their structure and behaviour so that each has its unique feeding niche.

Gulls

The generalists of the bird world, gulls feed upon a vast range of foods, from fish to earthworms, and carrion to domestic refuse. Their stout bills and robust digestion allows them to feed opportunistically, which makes many gull species highly adaptable. The name "seagull" is quite inappropriate: they are found almost everywhere, from city rooftops – in effect, artificial cliffs – to the open oceans.

FOOTPRINTS IN THE SAND
Until erased by the tide, soft sand and mud can reveal its use by birds, in the shape of footprints and pellets, the indigestible remains of meals.

grey or black above and white below, for camouflage against sea and sky

short, often powerful bill

webbed feet

black cap

long, pointed bill

short legs

long tail feathers

IN FLIGHT
A forked tail, long and pointed wings, and an elegant flight action have given rise to the alternative name for terns of "sea-swallows".

Terns

Built for precision flying, most terns feed by hovering, then plunge-diving on their fish prey, caught with the pointed bill. Food is carried back to the chicks until they can fly, so breeding colonies are close to rich feeding grounds. When not raising young they stay further out to sea. The arctic tern has the longest migration of any bird, flying an average of 2.4 million km – equivalent to three round trips to the Moon – in its lifetime.

Waders

A combination of long bill and long legs makes waders well adapted to feeding on invertebrates in soft mud. But the whole story is far more complex – different bill lengths and shapes give access to different foods, while longer legs allow feeding in deeper water. Waders congregate mostly outside the breeding season, and especially in high tide roosts when their feeding grounds are unavailable.

IN FOR THE KILL
The oystercatcher's blade-like bill can prise open bivalve molluscs such as cockles and mussels.

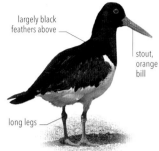

largely black feathers above

stout, orange bill

long legs

OYSTERCATCHERS

small black eye

long, pointed bill

strongly barred, brown plumage

sleek, slender body

straight bill

SANDPIPERS

large eye

head and breast markings

short bill

piebald plumage

very long legs

fine, upcurved bill

CURLEWS

long, stout legs

PLOVERS

AVOCETS

HOW WADERS FEED

The length of a wader's bill is a factor in determining its diet. Some use visual clues to find their prey, picking at surface food, turning over stones, or probing siphon holes in the sand. Others rely on touch, often using a regular "sewing-machine" action and specialized muscles that allow them to open the tip of the bill deep in the mud and capture their prey. Watch carefully – often the bird will bring its food to the surface to wash it, giving us the chance to see exactly what it is eating.

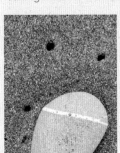

FEEDING HOLES

curlew   godwit   oystercatcher   sandpiper   plover   turnstone

crab

cockle

clam

snails

lugworm

NICHE FEEDERS
Bill shape and length, leg length, and feeding technique have evolved to give each wader family its own feeding niche, allowing outwardly similar species to co-exist in the shoreline habitat.

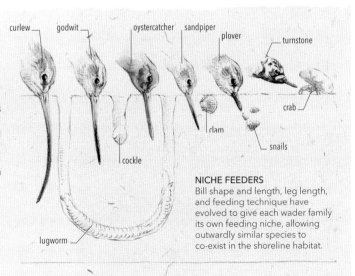

Seal colony

Although they spend a lot of time in the water, all seals need to come ashore to breed, and many gather in colonies.

Seals and sea lions both belong to a group of aquatic, warm-blooded mammals called pinnipeds. Pinnipeds live part of their lives in water and part on land – their flippers and torpedo-shaped bodies make them well-suited for diving and moving gracefully in water. The group is split into three families: walrus, eared seals, and true seals. True seals include the common (or harbour) seal, which is widespread in the northern hemisphere. In the North Atlantic, common seals often form mixed colonies with grey seals, however, the two species have different characteristics. Courtship and mating by common seals takes place in the water; they come ashore to rest and to pup. The pups can swim as soon as they are born, so breeding colonies can be on sand banks and flat beaches. In contrast, grey seal colonies are more active and mating takes place on land. Bulls fight to secure the best areas of beach and with them the most females. Their pups cannot swim for the first few weeks of their life, until the first white coat is shed, so they are born on rocky islets or pack-ice, above the reach of the tides.

VISITING A SEAL COLONY

Seal colonies can often be seen easily from land or sea, but care must be taken when viewing them not to cause disturbance. Pups may become separated from their mothers if a colony is spooked. Each species has a distinct breeding season – common seal pups are born during the summer months, while grey seal pups tend to be born in the winter. Timings vary across their geographical ranges however, so do take advice from local experts.

BOAT TRIP
Seal colonies provide the basis for many ecotourism initiatives. Approaching by boat allows close viewing with minimal disturbance.

GREY SEALS
Grey seal pups remain out of the reach of the tide for several weeks after birth, but storm waves may sometimes wash them out to sea.

Beach close-up

Walk along a tide-line anywhere in the world and you'll find the remains of coastal and marine fauna and flora, which have washed ashore or drifted from the ocean.

DRIFTWOOD

GOOSE-BARNACLE

Stones are smoothed by the sea, some may contain fossilized remains.

SCALLOP SHELL

SPONGE

PEBBLES

AMMONITE FOSSIL

Shells litter some beaches, many more may lie buried under sand.

WINKLE SHELL

MERMAID'S PURSE

SAND-DOLLAR

LIMPET SHELL

WHELK EGGS

MUSSEL SHELL

OYSTER SHELL

RAZOR-SHELL

STARFISH

BROWN SEAWEED

Seaweeds are torn from rocks by rough waves despite their strong holdfasts.

SEA BEAN

Partial or whole skeletons and cases of sea animals are washed ashore or dislodged from rockpools.

RED SEAWEED

GULL FEATHER

CUTTLEFISH SKELETON

GREEN SEAWEED

SEA-URCHIN TEST

CRAB

JELLYFISH

Sandy beach

Sand is a hostile habitat for many plants and animals. All dune inhabitants must have adaptations to cope with difficult conditions.

How a dune forms

Onshore winds pick dry sand from the beach and blow it inland. Out of reach of the tide, sand mounds can be colonized by drought-tolerant plants – their roots help to stabilize the surface, while their shoots interrupt the wind flow, leading to further build-up of sand. This process continues until dunes are formed, sometimes more than 100m (328ft) high in favoured locations.

SANDY HOME
Mounds of vegetation interspersed with bare sand create microhabitats in which many animals thrive.

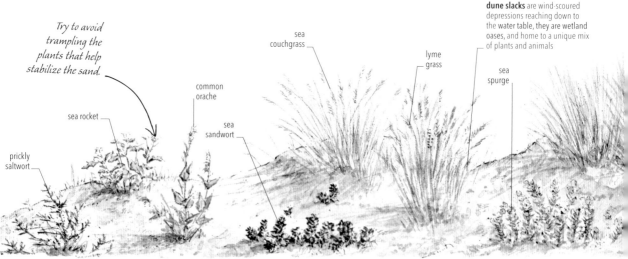

Try to avoid trampling the plants that help stabilize the sand.

prickly saltwort

sea rocket

common orache

sea sandwort

sea couchgrass

lyme grass

dune slacks are wind-scoured depressions reaching down to the water table, they are wetland oases, and home to a unique mix of plants and animals

sea spurge

Embryo dune

Low dunes on the seaward fringes are colonized by salt-adapted annual plants, which can complete their life cycles rapidly between the upheavals caused by storms.

PRICKLY SALTWORT

SEA ROCKET

Foredune

A short distance from the tideline, a range of creeping plants can get a roothold. Most of these have fleshy leaves for storing water in hot weather, and waxy leaf coatings or silvery hairs to reflect intense summer sunlight.

SEA SANDWORT

LYME GRASS

TURTLES

Ocean-wandering sea turtles come ashore on sandy beaches along some Mediterranean coasts to bury their eggs in deep holes, which they excavate in low dunes. Incubated by the warmth of the sun, the hatchlings emerge at night in an attempt to reach the relative safety of the sea before dawn brings the risk of predation.

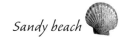

Dune builder

With its almost unlimited ability to grow both horizontally and vertically through depositing sand, marram grass forms the backbone of most large coastal dune systems. To survive drought, it has wax-coated leaves that roll up in dry weather, reducing the loss of water from its breathing holes on the upper surface of its blades.

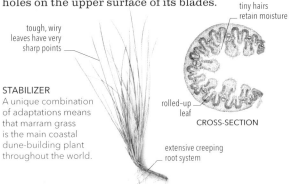

tough, wiry leaves have very sharp points

tiny hairs retain moisture

rolled-up leaf

CROSS-SECTION

STABILIZER
A unique combination of adaptations means that marram grass is the main coastal dune-building plant throughout the world.

extensive creeping root system

Dune dwellers

Unlike plants, animals can move or hide to avoid intense summer droughts. Reptiles, insects, and snails, for example, take refuge in the tussocks of marram grass, where they can take advantage of shade and trapped moisture. Amphibians, such as natterjack toads, bury themselves in moist sand, and sit out the drought until the rains return. Ghost crabs burrow all year round for safety.

DUNE DIGGER
Translucent ghost crabs inhabit deep burrows in sand, emerging at night to forage safe from gulls.

NATTERJACK TOAD

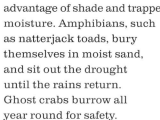

ragwort

marram grass

wild thyme

sand fescue

sea bindweed

Yellow dune

Marram grass adds stability to foredunes, promoting further dune growth. Other largely drought-tolerant plants follow, but yellow dunes still have a high proportion of bare sand. A lack of organic matter creates the sand's colour.

RAGWORT

SEA SPURGE

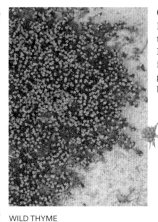

WILD THYME

Grey dune

More mature dunes are stable enough to support a greater diversity of plants. Dead leaves and other organic matter incorporated into the sand give it a greyish colour, which is often enhanced by extensive patches of lichen growth.

SEA HOLLY

Limestone

Usually pale grey or yellowish in colour, limestone is a very variable rock – a result of the variety of ways in which it was formed. Although often hard, producing erosion structures such as platforms, it is vulnerable to weathering by water and acid rain. Many types of limestone contain the fossilized remains of animals that inhabited the prehistoric seas when the rocks were formed.

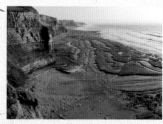

WAVE-CUT PLATFORM

LIMESTONE

Cliffs

Wherever rocks meet the sea, cliffs evolve. Their size and slope are dictated by the rock type. Erosion features such as caves, arches, and stacks reflect weaknesses in the rock that are more vulnerable to wave attack.

Chalk

A pure form of limestone, chalk is generally gleaming white and relatively soft. Under attack from the sea, it usually erodes into near-vertical cliffs. Each layer of the rock becomes visible as a ledge, which is often used as a nesting site by seabirds. In a chalk cliff, look for seams of flints. These glassy nodules formed the basis of the earliest human industry – the shaping of stone tools.

WHITE CHALK

NATURAL ARCH

Granite

Granite was formed by the cooling of molten volcanic rocks beneath the Earth's surface. This extremely hard, crystalline rock contains minerals that give it distinctive colours. Due to its make-up, it erodes very slowly, which is why granite cliffs are often stepped and well-vegetated – usually with few sheer drops and littered with large, rounded boulders.

WHITE GRANITE

STACK

Sandstone

A common sedimentary rock in which grains of sand are visible, sandstone ranges in colour from pale whitish to red, or even greenish. Sandstone cliffs often show layers that allow you to track environmental conditions present during deposition. Erosion acting upon these features may create many natural sculptures, including caves.

SANDSTONE CAVE

RED SANDSTONE

Volcanic

Throughout Earth's history, volcanic lava flows have solidified into a range of blackish rock types, especially basalt. The crumbly (friable) rocks in places of recent volcanic activity, such as the Canary Islands and Iceland, form some of the most impressive, barren cliff landscapes in the world. In such places the first stages of colonization by flora and fauna are visible.

BLOW HOLE

VOLCANIC ROCK

Look behind rocks or shrubs for more delicate plants such as maidenhair ferns, which take advantage of the natural shelter.

MAIDENHAIR FERN

SPRING SQUILL

SIX SPOT BURNET MOTH

Cliff view

Carved and moulded by wind and salt spray, the plants and animals that inhabit a clifftop can be as dramatic and beautiful as the steep, rocky slopes themselves.

Crashing waves and the cries of seabirds lend an air of wildness to cliffs, where life clings to a precarious existence in the face of often harsh elements. Revel in the grandeur by all means, but don't overlook the small stuff. Take time to look at the windswept summer turf and you'll see that it is studded with an array of beautiful plants and miniature flowers, which in turn host an array of insect life.

Look out for a frothy mass of plant sap — or "cuckoo-spit" — on cliff plants and you'll find froghopper nymphs inside. They secrete the sticky stuff to hide in for protection.

ROCK SAMPHIRE

CUCKOO-SPIT ON PLANT IN BUD

Holes in leaves are a sign
of plant-loving insects.
Turn over the leaves
to find butterfly and
moth caterpillars.

WILD CABBAGE
LEAVES

GOLDEN
SAMPHIRE

Bees and wasps visit
nectar-rich cliff flowers
in the summer. Look
for their nesting
burrows in sun-warmed
patches of bare soil.

SEA
MAYWEED

BURROWING
WASP

Cliff colony

Sea cliffs, especially in northern temperate and Arctic regions, are often home to large colonies of sea birds. They nest here as they are relatively safe from predators.

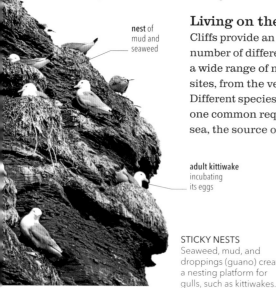

nest of mud and seaweed

adult kittiwake incubating its eggs

Living on the edge

Cliffs provide an excellent location for spotting a large number of different sea birds in summer. They provide a wide range of niches for sea birds to use as breeding sites, from the very top to just above tide level. Different species favour different places, but all share one common requirement: the need to be close to the sea, the source of much – if not all – of their food.

POINTED EGGS
Long thought to prevent them rolling off ledges, the pointed shape of guillemot eggs actually improves their ability to stand up to impacts from crash-landing adults.

pointed tip

STICKY NESTS
Seaweed, mud, and droppings (guano) create a nesting platform for gulls, such as kittiwakes.

GANNET COLONY
Gannets live in large colonies. The safest spots are in the centre – younger birds nest at the edges where gulls take eggs and chicks.

Sea watching

Cliff tops provide an excellent vantage point for scanning the ocean for birds and marine mammals, such as porpoises. Watching the sea takes patience as for much of the time there may be little to see. Make sure you have some shelter, warm clothing, and good binoculars. The rewards will come when you spot something, perhaps a frenzy of gannets diving on a shoal of fish or a pod of whales or dolphins passing by.

COASTAL VISITORS
Promontories such as cliffs give you the chance to watch dolphins as they pass close to shore.

Watch birds from a boat to avoid disturbing them.

BIRDWATCHING
Viewing cliff birds requires great care. Always ensure you remain behind any safety fences or, better still, take a boat trip to view the spectacle from below.

What nests where

SKUAS
Skuas nest on the ground on cliff tops, close to other birds from which they steal food and often their chicks.

KEEPING COUNT

Monitoring seabird colonies helps to measure the health of the marine environment; photographs can record changes. Recent declines in many areas have raised issues of overfishing, pollution, and climate change. Accurate counts are tricky, especially when some birds are away feeding in open water.

PUFFINS
Although their numbers are in decline in places because of falling populations of their sandeel prey, puffins are easy to spot with their colourful bills. They dig burrows in soil on upper slopes and cliff tops.

GUILLEMOTS
Lining the ledges, in close-knit ranks, guillemots and razorbills are among the most numerous of all the cliff-nesting sea birds. Both species dive underwater to catch their small fish prey.

EUROPEAN SHAG
The large, untidy nests of shags are often found in caves or crevices near the foot of a cliff, but are sometimes much higher. White droppings reveal their location.

Cliff close-up

The wildlife found on cliffs varies depending on location and the geological make-up of the habitat. Most cliffs will have low, matted plantlife, while sandy cliffs facing the sun are rich in insects.

SEA BEET

Clifftop flowers often grow in mats to reduce exposure to high winds.

STONECROP

ROSEROOT

TREE MALLOW

BUCK'S HORN PLANTAIN

ROCK SEA LAVENDER

COMMON BLUE
BUTTERFLY

PROVENCE
BURNET
MOTH

RED
ANT

*Insects and snails feed
on plants and some
make their homes in
the loose soil.*

LICHEN-COVERED
ROCKS

KIDNEY
VETCH

*Lichen thrives on
rock faces exposed
to sunlight and
sea spray.*

BANDED
SNAIL

CLADONIA
LICHEN

*Rocks can reveal
the remains of
sea life that
lived thousands
of years ago.*

CARLINE
THISTLE

COLT'S
FOOT

FOSSILS IN
LIMESTONE

BIRD'S FOOT
TREFOIL

AMMONITE
FOSSILS

THRIFT

Estuary

Where a river meets the sea, the resulting estuary becomes a fluctuating mix of fresh and salt water. Interlaced with a range of other wetland types, such as salt marshes and mudflats, this mosaic of shallow, open water channels can harbour underwater sea-grass meadows – food for water birds and vital habitats for creatures such as sea horses and other fish.

COMMON
SEA HORSE

EEL
GRASS

GREY MULLET

Coastal wetlands

On low-lying coastlines, the boundary between land and sea is blurred by the presence of coastal wetlands. Influenced by salt water and tidal movements, they make up some of the richest wildlife habitats in the world.

Salt marsh

Where a mudflat surface is exposed for a sufficient part of the tidal cycle, salt-tolerant plants take hold to form salt marshes, providing swathes of colour as they bloom during the summer months. Due to their deep tidal creeks, they aren't easily accessible to people, which adds to their attraction to birds, such as gulls, and other wildlife species.

BLACK-HEADED GULL

SEA
PURSLANE

SEA LAVENDER

Mudflat

Extensive mudflats, washed by every tide, are at the heart of many coastal wetlands, and are home to a vast range of invertebrates, including molluscs such as cockles and clams. These in turn attract wading birds such as sandpipers when the tide is out, as well as fish when the tide floods in. Largely featureless, apart from shallow creeks and pools, mudflats play host to some remarkable concentrations of species on a global scale.

CURLEW SANDPIPER

COCKLE SHELL

Salinas

Around Mediterranean coasts, artificial lagoons were created to obtain salt from evaporating sea water. These salinas provide an important habitat for many herons, shorebirds, and gulls. Shorebirds such as dunlin, sanderling, and curlew sandpipers search for invertebrates in the shallows, while black-winged stilts, little egrets, and greater flamingos wade in deeper water. Reptiles, including Montpelier snakes, live around the margins.

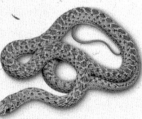

MONTPELIER SNAKE

GREATER FLAMINGO

Tidal marshes

Washed by the highest tides, salt marshes are dynamic habitats of low-lying coastlines, abounding in specialized plants and animals.

Salt-marsh strategies

Protecting the land by absorbing the energy of the sea like a sponge, salt marshes are nature's own sea defences. All the plants that make up these marshes must be able to thrive in salty water. Many have desalination cells, which strip salt from the water, leaving fresh water for the plant's use. Other adaptations in these plants include some way to get rid of excess salt – you can see the crystals of excreted salt on their leaves – and succulent tissues in which to store available fresh water.

MARSHLAND SAFETY

Deep creeks and pools, soft mud, and the relentless tides mean that great care is needed if you explore a salt marsh. Luckily, salt marshes are flat, so much of their fascinating wildlife can be viewed from the safety of nearby higher ground – binoculars are essential.

STAND TOGETHER
Local knowledge is invaluable. Walking with a guide is the safest way to explore marshes.

Submergence marsh

Each tide brings in a fresh supply of silt. As this silt is deposited, the mud surface rises, and eventually plants begin to germinate and colonize. The lower submergence marshes are washed by every tide, but plants like cord grass and sea aster have air spaces in their tissues that allow them to survive the time they spend underwater. Animals here include fish and crabs, which take advantage of the rich food supplies.

EUROPEAN BASS
Salt marshes support spawning and nursery areas for many fish species, such as European bass, which move in at high tide.

SHORE CRAB
Marine creatures such as crabs remain in the marsh at low tide, taking refuge in pools or burying themselves in the mud.

Emergence marsh

Mid-level marshes are covered by the higher tides of the monthly cycle (see pp.180–81), so salt is a an ever-present challenge, which is why salt marshes are rarely as diverse as their freshwater counterparts. But many salt-marsh plants have attractive summer flowers, such as the vibrant, purple sea lavender, and these in turn attract insects.

GROUND BEETLE
Some insects such as beetles can live in leaf litter on the marsh surface.

Salt-marsh birds

Regular soaking by the tides prevents birds from breeding on all but the highest-level marshes. Above the reach of summer tides, however, waders and gulls nest, sometimes in large colonies. In winter, many species of birds use all areas of the salt marsh – which makes it a great place to view them from a safe distance in a hide. Waders roost on the high marshes, safe from ground predators, and feed in the muddy creeks; ducks and geese graze at all states of the tide; and large flocks of finches and buntings make use of the abundant supply of seeds produced by the salt-marsh plants.

MARSH DABBLER
Many ducks, such as this teal, head for salt marshes in winter, where they graze and dabble for the nutritious, oil-rich seeds.

WETLAND FISHER
Even at the lowest tide, the network of creeks and pools in a salt marsh provide very rich pickings for birds like herons and egrets, such as this great white egret, which feeds on crustaceans and fish.

ARROW GRASS
The fleshy leaves of sea arrow grass can be distinguished from true grasses by their sweet, aromatic scent when crushed.

Upper marsh

Above the level of all but the highest tides, the upper marshes are often dominated by low shrubs, typically members of the spinach family. Infrequent tides followed by evaporation in sunlight can produce extremely high salt concentrations, however, so most plants have fleshy leaves that can store rainwater and help to buffer the effects of salt at their roots. Insects such as crickets are common here.

MARSH MALLOW
At home in salt or fresh water, the mallow can be found in damp marshes and tidal river banks.

BUSH CRICKET
Safe from the risk of frequent flooding, a wide range of insects can be found in the upper marsh zone. Some, like the bush cricket, are almost invisible in the green foliage.

Mudflats

Found in sheltered areas, such as estuaries, mudflats are made up of very fine particles of material deposited by the sea and river water.

MUDDY MOSAIC
Estuarine mudflats form an intricate mosaic with water channels and salt marsh.

How mudflats form

As soon as the fine particles, or silt (see p.182), settle out, plants begin to colonize it. Firstly, microscopic algae called diatoms start to grow on the surface, helping to "glue" the silting together and make it more stable. The seeds of salt-tolerant plants, carried by the tides, can then germinate – their roots give even more stability to the mud and their shoots help to slow down water movement, which means that even more silt is deposited when the sea washes over it.

LIFE IN SALT WATER
Glasswort has a range of adaptations for life in salt water, including fleshy, cylindrical stems for storing fresh water.

What to spot

At first glance, mudflats may seem bleak and devoid of life, but take a closer look and you will notice an abundance of marine creatures and plants. This rich habitat is reliant on twice-daily tides that supply the mudflats with food as well as fresh silt. Even when the wildlife is not visible, it is often possible to see evidence of their activities, such as worm casts and feeding tracks. If you decide to visit a mudflat, keep an eye on the sea and be careful not to get cut off by the rising tide (see pp.180–81).

1 Gulls and other water birds that feed on the flats leave their tracks in soft mud.

2 A gull's pellet is the indigestible remains of its last meal – handle with care and be sure to wash your hands after touching it.

3 Cockles and other filter-feeding shellfish form dense beds on the surface or in the top layer of mud.

4 In shallow water, fan-worms extend a crown of feathery feeding tentacles from their rubbery mud tubes.

5 The shells of mussels are attached by strong threads to stones and other hard structures that are buried in the mudflats.

6 Dog-whelks leave feeding tracks as they move over the surface, feeding on dead organic material.

7 Algal mats provide food for grazing wildfowl and snails.

8 Mud-snails are tiny, but numerous and provide miniature morsels of food for throngs of wading birds.

9 A shallow depression marks the inhalent end of a lugworm's u-shaped tube, through which it draws water and food.

10 Lugworms feed on organic matter within mud and sand, excreting the indigestible remnants as worm casts.

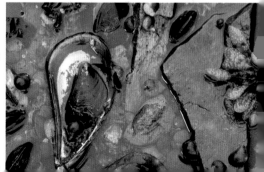

What lies beneath

To get a true picture of mudflat life, you cannot just look at the surface. Buried in the mud, sometimes at considerable depth, is a range of burrowing and tube-dwelling invertebrates. Some filter food from the sediment, while others prey on the animals around them. They stay in their burrows, where they are protected from drying out when the tide recedes, and only show themselves at the surface of the mud when it is covered with water. At high tide, they are preyed upon by fish, but at low tide wading birds take their toll. The length of wading bird bills affects the range of prey available to them (see pp.186–87).

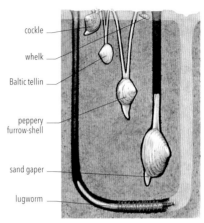

cockle

whelk

Baltic tellin

peppery furrow-shell

sand gaper

lugworm

BURROWERS
Each type of burrowing animal lives at a particular depth within the mud.

PEACOCK WORM

RAGWORM

BOOTLACE WORM

BURROWING MUDSHRIMP

MUDDY DWELLERS
Whether pushing through the mud or inhabiting a tube, the hidden riches of a mudflat add three-dimensional complexity to a hugely diverse ecosystem, as productive as any other habitat on Earth.

SIEVING FOR LIFE

To fully appreciate the richness of mudflat life, you need to get your hands dirty. Dig up a small sample of mud with a trowel and put it through a series of sieves, starting with the largest mesh. This will retain the larger shells and lugworms, while a finer mesh will hold back mud-snails and smaller worms. Even mud that has passed through both sieves will still have life in it, including larvae and nematodes that are only visible through a microscope.

CLEAN SWEEP
A garden riddle, sieve, and brushes are the basic tools to tease apart any specimens.

All at sea

Most of the oceans' depths have never been explored, but you can watch many species from the surface of Earth's last great wilderness.

Exploring the big blue

Over 70 per cent of the world's surface is covered by seas and oceans, which have an average depth of over 3km (2 miles). Most ocean depths are beyond the reach of all but the most specialized submarine, but you can explore coastal seas – both above and below the waves – with relative ease. Many types of animal life are within easy reach of the seashore, and taking a boat trip can provide an insight into the lives of seabirds, seals, whales and dolphins, and other mammals such as sea lions and sea otters. Underwater, diving or snorkelling can bring you into contact with another experience of ocean wildlife, and give you a totally different, more intimate encounter with many types of marine animals.

SWIM WITH DOLPHINS
Dolphins are naturally inquisitive and highly intelligent mammals, and may approach visitors in their environment.

BOAT WITH A VIEW
Take a ride in a glass-bottomed boat in areas with clear seas, and you can get a close-up view of sea life.

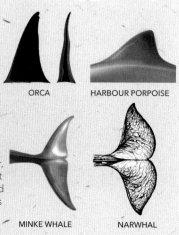
IDENTIFYING FINS AND TAILS
The shape and marks on the dorsal fins of cetaceans allow scientists to recognize individual animals. The challenge is to identify species by fin profile as they surface to breathe. The tall, black dorsal fin of an orca, curved in females, is unmistakable, as is the curved back of a minke whale and its tail silhouette before a dive. Porpoise sightings, however, are more fleeting; look for a straight leading edge to their dorsals instead of the curve of a dolphin's. Narwhals lack a dorsal fin, but their tails have a distinct notch in the middle.

ORCA

HARBOUR PORPOISE

MINKE WHALE

NARWHAL

Taking the bait

Many marine animals rely on speed to escape predators, but small fish such as anchovies or sardines gather in shoals for protection. Large shoals can also form at an upwelling, where smaller fish gather to feed on plankton. Just like large flocks of birds, a shoal of fish can move as one, whirling and changing direction so rapidly that it is difficult for a predator to pick off any one fish. However, when many predators come together, the odds change, and a wildlife spectacle unfolds as the shoal changes from a safe haven into a "bait ball". In this situation, attackers strike from all sides; the fish are driven towards the surface by dolphins, whales, or sharks, where they are picked off by waiting seabirds, such as gannets.

DIVING GANNETS
Plummeting from a great height, gannets can penetrate deep into the water to spear their prey.

BAIT BALL
Clumps of fish such as sardines attract predators from above and below – sharks, tuna, whales, and diving birds.

tubular nostril above bill   short tail

STRANGER TO THE GROUND
Shearwaters spend almost their entire lives at sea, flying thousands of kilometres every year.

SAVE OUR SEAS

Modern fishing methods, climate change, and pollution are stripping oceans of life and upsetting their natural balance. We can take practical steps to lessen our impact on the oceans by keeping beaches clean, supporting marine sanctuaries, and safely disposing of harmful pollution.

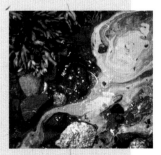

POISONOUS OIL
Oil spills destroy the waterproofing on seabird feathers and the fur of seals and otters, then poisons them as they preen or clean themselves. It also affects the entire food chain.

DANGEROUS RUBBISH
Any rubbish in the sea such as the discarded fishing gear above, is gathered by currents. The North Pacific Garbage Patch, between Hawaii and California, contains millions of tonnes of toxic plastic.

Tundra and ice

Freezing temperatures present life with problems. Combine these with high altitudes, extreme weather, or long periods without sunlight, and you have one of the most challenging environments on Earth. Yet if there are resources available, no matter how sparse or apparently remote, life will reach for them and adapt to access them. The barriers of glacial cold, unremitting darkness, or thin air have been breached by a guild of specialists, which thrive where we would shiver to a fatal standstill. And of course, the fate of these remarkable communities is now under serious scrutiny as these fragile habitats succumb to climate change.

The Arctic tundra

The treeless Arctic tundra is a great place to see wildlife such as reindeer, musk ox, and grey wolves from the comfort of a guided, specialist tour.

What is the tundra?

The Arctic tundra is a vast landscape north of the tree line (see pp.144–45), extending across northern Scandinavia, Russia, Siberia, Alaska, and Canada. It is a habitat shaped by extreme cold. For much of the year, it is snow-covered, dark, and windy. Soil in this region is almost perpetually frozen and is called permafrost. This limits the growth of roots, so the only plants that survive there are small shrubs, mosses, and lichens. In summer the upper permafrost melts, transforming the tundra into a marshy bog that supports a host of wildlife.

SOLID GROUND
Permafrost is soil that remains below the freezing point of water. Plant life blooms in summer when its upper layer thaws.

What you might encounter

Since few animals can tolerate its harsh, cold conditions, the tundra is a place of low biodiversity. However, you may see reindeer and musk ox grazing on small plants and lichen. Predators include Arctic foxes and wolverines, and smaller animals include Arctic hares and lemmings.

ICE SURVIVOR
Lemmings survive the cold by burrowing underground. They migrate when their numbers swell, but many drown in rivers and lakes.

SUMMER VISITOR
The polar bear isn't just a creature of the ice. In summer they move into the Arctic tundra, where they may be seen on a wildlife tour on the island of Spitsbergen.

THREATS TO THE TUNDRA

Tundra all over the world is under threat. Mining, drilling for oil, and new pipelines such as that from Russia's Yamal Peninsula directly impact the environment, but the most severe threat is global warming. Polar temperatures are rising faster than average. This melts permafrost and releases greenhouse gases such as carbon and methane, which could have damaging results for the planet as a whole.

A RUSSIAN GAS PIPELINE

LICHEN GRAZERS
Reindeer (called caribou in North America) spend the summer on tundra. In winter they migrate to graze in forested areas, feeding on lichens, and grasses.

What to see in the summer

Winters are cold, windy, and harsh, but visit the tundra in summer and you'll see a transformed landscape. Long days of almost 24-hour sunlight warm the topsoil, melting the surface layers and turning the environment into a lush, boggy marshland where many plant species can grow. Summer is a good time to see animals that migrate to this region; they do so to avoid predators as well as to feed on abundant insects and fish. Reindeer are just some of the migrants, roaming many hundreds of kilometres to graze on the summer plant life. You can also see bird species such as brent and red-breasted geese, and red-necked phalaropes, which use the marshlands as a place to raise chicks.

SPIN FEEDER
A red-necked phalarope's red collar is clearly visible when it flies. When feeding in shallow water, these birds spin in tight circles to create upwellings of food.

SILENT HUNTER
Snowy owls prey on small animals such as lemmings. With no trees available in the tundra, the owl makes its nests on the ground.

SUN FOLLOWER
The Arctic poppy is a miniature version of its relative in temperate regions. Its tiny flowers turn their heads to follow the Sun.

Even the snowy owl's huge feet are feathered to help it cope with the cold.

WATCH OUT FOR MOSQUITOES

We may think of mosquitoes as tropical insects, but the summer tundra teems with them. Because tundra is flat with frozen permafrost below, meltwater from the surface has nowhere to go. Stagnant puddles are warmed by 24-hour sunlight, making them ideal for mosquito larvae. This is good news for waterfowl, which feed on the larvae, but less good for reindeer and human visitors, who are plagued by the blood-sucking insects. Take your insect repellent!

Arctic fox

In the tundra the change between seasons is extreme and the animals that live there must adapt to survive.

The Arctic fox lives in some of the coldest parts of the planet. The fox stores heat within its body by exposing less of its surface to the cold, with stout legs, a short muzzle, and small rounded ears. Its chief adaptation for dealing with the cold in the icy Arctic winters is its fur. The Arctic fox is the only member of the dog family to change the colour of its coat with the seasons. In spring, the fox is tawny brown but as winter comes, thick white hair grows through. It has one of the densest fur coats and the hairs of its winter coat are almost double the length of those of its coat in summer, and the thick, deep fur provides warmth. Every part of its body is covered in fur – even the pads on the soles of its feet, helping it to walk on ice. The reason for the fox's colour change is to blend in with the white of the environment. This allows it to sneak up on its prey and avoid larger predators. The Arctic fox preys on small rodents, such as lemmings. It has such sharp hearing that it can hear them rustling and pounces on them through the snow.

SEASONAL FUR

Both hunter and hunted employ similar strategies to avoid detection. Arctic foxes prey on Arctic hares, although their large size makes them intimidating game. By blending in with its surroundings the fox can use stealth to approach its quarry. The hare's white colour helps it avoid its predators, which also include Arctic wolves and snowy owls.

CAMOUFLAGE COLOURS
The Arctic hare in winter (left) and summer (right). In winter its white coat usually blends with snow to avoid the eyes of predators.

HIDE AND SEEK
In winter the thick coat of the Arctic fox turns white, blending with the snow and ice. This camouflage helps it to sneak up on rodents, birds, and occasionally ringed seal pups.

Life on the ice

The Arctic is among Earth's last wildernesses. Although Arctic areas are changing fast, wildlife thrives there.

DISTANCE FLYER
Arctic terns winter in Antarctica, but breed in the Arctic, so may travel up to 40,000km a year.

Visiting the ice

Visiting a polar region can be the most exciting trip you ever make, yet due to the sensitive nature of these fragile environments, tourists must respect them. Many Arctic destinations, such as the Norwegian island of Spitsbergen, are best seen from the water; the best time to visit is from May to September. Remember that these are remote, pristine, and extreme locations. Listen to your guide, respect the animals and the ice, and you can have some of the most memorable wildlife encounters of your life.

What you will see

In the Arctic, expect to see whales, walruses, and, if you're lucky, polar bears. While the last are top predators with little fear of people, cautious encounters from a safe distance can be magical. Ivory gulls feed on any carrion they leave behind. With luck, you may see a beluga whale or a narwhal around the edge of the ice, where auks and other seabirds dive for fish.

What to pack

1. Binoculars, for spotting birds and whales.

2. Sunscreen to protect your skin – the ozone layer is thin at the poles.

3. Sunglasses to cut glare reflected from the ice.

4. Camera to record what you see.

5. Warm clothing, especially for hands and feet.

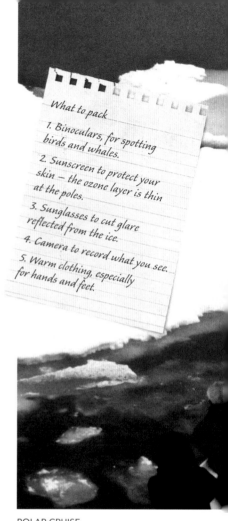

POLAR CRUISE
Arctic cruise ships with toughened, ice-breaking hulls sometimes encounter a polar bear.

LIFE UNDER THE ICE

In the Arctic Ocean, the foundation of the food chain is made up of billions of tiny crustaceans called copepods and krill, which feed on even tinier organisms under and around the ice. They provide food for sea birds, seals, and whales.

KRILL

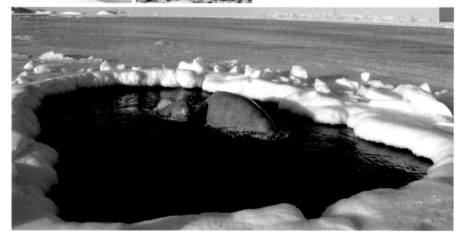

1 Polar bears are powerful predators, at home on ice and in water, where they are superb swimmers. They can eat prey as large as a walrus.

2 Large, flippered marine mammals, walruses have a thick layer of fat, called blubber, under their skin to keep them warm in icy water.

3 Belugas are one of the so-called Arctic "ice whales". They travel long distances beneath the ice, using holes and cracks to surface and breathe.

EARTH'S VITAL ICE

Climate change is drastically affecting polar regions, which must be conserved for all species – not just those that live there. Average Arctic temperatures have risen at twice the rate of those elsewhere, while the Antarctic Peninsula has risen to five times the average. This has profound implications for the planet. An increase in the Earth's temperature causes icebergs to melt, raising sea levels. The Greenland ice sheet is the area scientists consider most at risk: if it melts, sea levels could rise by up to 7m, bringing flooding to many coastal areas.

Scientists have shown that little auks' diet has changed as Arctic sea ice has retreated.

Ice structures

Ice occurs in very different forms on land and in the ocean. On land, centuries and millennia of snowfall, layer upon layer, builds into glaciers and ice sheets that can be kilometres thick. In the Arctic, ice sheets and glaciers can flow directly into the sea, where they "calve" icebergs – floating mountains of ice. Ice also forms when the sea itself freezes, but this results in much less substantial "sea-ice". Sea-ice is only up to a few metres thick and is often much less. Sea-ice attached to land is called "fast ice" while "pack ice" floats freely, carried by wind and currents.

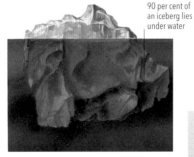

90 per cent of an iceberg lies under water

HOW ICEBERGS FLOAT
Ice floats because it is less dense than water, thanks to open spaces formed between the hydrogen molecules when water is frozen.

MELTING ICE
Pack ice (above, right) refers to a body of drifting ice, which is carried along by wind and surface currents. Icebergs (below) are carved by the sea and sun into various organic shapes and sizes.

Glossary

Abdomen
In mammals, the part of the body between the thorax (chest) and pelvis; in insects, the hind section of the body behind the thorax.

Adaptation
The evolution of features within a species or population, improving fitness for life in a particular habitat.

Agro-forestry
Cultivation of trees as a crop, for timber, pulp, or other products, such as palm oil.

Algae
Simple organisms, the most complex being seaweeds, lacking structures of plants such as roots and leaves.

Arctic
The region around the North Pole, north of the Arctic Circle at 66° 33'N.

Biodiversity
The total variety of living things, including species and subspecies.

Biome
An ecosystem, or community of plants and animals, living in particular geographical and climatic conditions.

Botany
A branch of biology specializing in the study of plants.

Cambium
In plants, a layer of cells within stems and roots whose growth gives an increase in girth.

Chlorophyll
Green pigment in plants, vital in extracting energy from sunlight.

Chrysalis
Pupa, a life stage in insects and moths, between larva and adult.

Climate change
A gradual change of global climatic balance, natural or artificially induced.

Cocoon
A silky protective casing produced by some caterpillars.

Colony
A grouping of breeding animals in a specific site, for social stimulation or protection.

Commensal
A lifestyle relationship between two organisms, to the benefit of one with no harm to the other.

Compound leaf
Leaf split into several leaflets.

Contour
Line on map (or, imaginary, on ground) joining points of the same altitude.

Cyclone
An area of winds rotating inwards to "fill" a central area of low pressure; also a name for tropical storms in the Pacific and Indian oceans.

Dabbling
The taking of water, debris, seeds, and tiny organisms into its bill by a surface-feeding duck; water is expelled with its tongue, and food is retained.

Debris
Assorted mixture of material washed or fallen from above: from hillside rocks to fine soil and plant material in tree bark cavities.

Deposition
The laying down of suspended items washed along in a current.

Dorsal
On the upper part of the body; view from above.

Ecosystem
Complete assemblage of living things, from soil organisms to higher plants and animals, living in particular conditions and geographical area.

Enzyme
Protein produced by living organism, helping to speed up chemical reactions.

Epidermis
Outer layers of the skin.

Epiphyte
A plant growing on another plant without parasitizing it.

Evergreen
Having leaves all year, which are shed and replaced more or less continually, not seasonally.

Fern
A plant with vascular system, roots, leaves, and stems, but reproducing by spores instead of flowers.

Filter-feeder
An animal that takes a mouthful of water containing minute food and expels it through a filter, retaining the food, for example, whales with "whalebone" or baleen plates instead of teeth.

Fragmentation
Past extensive distribution of plant or animal, now reduced to small, isolated or remnant areas, through climate change or human action.

Fresh water
Water from rain, in lakes, rivers, marshes, and aquifers, with low concentration of dissolved salts and minerals (these increase through brackish to salt or sea water).

Friable
Crumbly and easily broken down.

Fungus
Plant-like organism that does not create its own food with chlorophyll, lacking any green pigment, typically feeding on remains of dead plants and animals.

Gall
Growth on plant leaf or twig in response to attack from parasitic insect, mite, fungus, or bacteria; parasite often identifiable by particular shape and colour of gall.

Genus
A unit in scientific classification of living things, linking similar species – first of two-word "scientific name", for example, *Homo* in *Homo sapiens*.

Germination
Period when seed or seedling emerges from dormant period, such as winter, to begin growth.

Gill
A structure that extracts oxygen from water, in fish or early stages of amphibian; also structure beneath cap of some fungi, containing spores.

Glacier
A mass of ice that becomes so heavy that it gradually "flows" imperceptibly downhill.

Greenhouse gas
Gas, such as carbon dioxide, that allows heat from the Sun to reach the Earth, but prevents it from radiating outwards, hence increasing global temperature.

Gyre
A large-scale circulation of ocean surface currents.

Habitat
The amalgamation of features, such as soil, plants, animals, and local climate, in which a particular organism lives.

Harem
A group of females assembled and defended for reproductive purposes by one male.

Heliotropism
Movement of plant during the day, "following" the movement of the Sun.

Herbaceous
Plant that dies back to soil level in autumn and winter.

Hibernaculum
A structure made to give a safe site for hibernating reptiles and amphibians in winter.

Humidity
The amount of water vapour in the air.

Humus
Decaying vegetable matter in the upper layer of soil, giving it a dark brown or black colour.

Hyphae
Long filaments of fungi, on or below ground, that form the mycelium; extracts and transports nutrients.

Lateral line
A line of sensitive cells along the side of a fish, able to detect sound and movement.

Leaflet
Division of a compound leaf, such as an ash leaf.

Lek
Communal display of males of some birds, such as black grouse and ruff, to attract and impress females; also the name for the traditional site used for such displays.

Lichen
Organism formed by close liaison of a fungus and a green plant that takes energy from sunlight, such as a green alga.

Litter
Fallen leaves collecting beneath trees and shrubs, decomposing over several months.

Mantle
On a bird, feathers cloaking the upper part of the body; a bird of prey also protects its catch by "mantling", opening its wings over its food.

Meander
Wide, S-shaped bend or loop in a river; fast flow undercuts the outer edge of a bend while slower flow deposits shingle on the inner edge, gradually shifting a meander downstream.

Melanin
Dark pigment, for example, in fur and feathers, giving darker, richer colours and black, and also adding strength to colour.

Metabolism
The sum of all physical processes that take place in the body.

Metamorphosis
Marked and rapid change between life forms of certain groups of animals, for example, from caterpillar to chrysalis to butterfly.

Microbe
Microscopic organism, or micro-organism, almost invisible to the naked eye.

Midrib
Central stiff support of flat leaf.

Migration
A regular, often annual, large-scale movement of animals of a particular species, such as wildebeest and swallows, often in connection to seasonal changes in climate and food, or for breeding purposes.

Monsoon
Seasonal wind and associated rainfall, producing majority of annual rain in one short season: West African and Asian–Australian monsoon systems are the biggest.

Mycelium
Network of fibrous filaments, or hyphae, beneath a fungus, which collects nutrients.

Mycorrhiza
A close association between a fungus and roots of a plant, to the benefit of both.

Native
An organism in its natural geographical range, i.e. not introduced, either deliberately or accidentally, by human action.

Nymph
Stage in the life of some insects, looks much more like the final adult form than a typical larva.

Opposite
Describes leaves or leaflets arranged in opposing pairs on a stem.

Outer skeleton
A shell-like, structural outer layer of certain invertebrates.

Oxygen
Abundant, tasteless, colourless gas in the atmosphere, essential to life; also in water and other natural substances.

Palmate
Having a web-shaped form between "fingers", for example, the feet of a palmate newt, or the shape of a sycamore leaf.

Permafrost
Permanently frozen soil, often causing waterlogged ground when higher layers thaw out in summer.

Pheromones
A chemical signal between insects, for example, laying a "food trail" or attracting a potential mate, often over remarkably long distances.

Photosynthesis
Extraction of energy from sunlight by chlorophyll in plants, and conversion to sugars and carbohydrates.

Pigment
A chemical material that influences the colour of reflected light by absorbing various wavelengths.

Plankton
Assorted minute plants, animals, and bacteria living and drifting freely in upper layers of water.

Polar
An area close to the pole: an imprecise definition, but closer to the pole than "Arctic" or "Antarctic".

Pollination
Fertilization of plants as male pollen grains are transferred (by wind, insect, or bird) to female reproductive structures.

Precipitation
Water vapour coagulating in the atmosphere as its capacity to absorb water reduces through changing pressure or temperature, to form rain, sleet, hail, or snow.

Prehensile
Mobile or capable of grasping, for example a prehensile tail that can be curled to grasp a branch.

Proboscis
Elongated structure from an animal's head, especially a tubular probe from an insect.

Pupation
Period in metamorphosis of some insects in which larval structures break down and adult features develop.

Rut
Period when males (for example, of deer) gather females into groups for reproduction, and to defend them against other males.

Saliva
A secretion from the mouth, serving as lubrication for swallowing food and also as "glue" to help create external structures, such as nests.

Scale
A small, rigid plate-like structure growing from the skin, for protection and colour, for example, on fish or butterfly.

Sciophyte
A plant that can thrive in shaded areas.

Sediment
Particles initially suspended in water, deposited as water velocity reduces or particles coagulate into heavier items.

Shoaling
Fish living in groups, with a degree of collective action.

Species
A basic unit in the classification of living things that groups together genetically similar individuals: members of a species interbreed and produce fertile offspring recognizably of the same species.

Spinneret
Organ of a spider that spins its silk fibre or web.

Stipe
The stem of a typically toadstool shaped fungus.

Substrate
Underlying rock or subsoil beneath the soil.

Symbiosis
Arrangement in which two or more organisms live inextricably and closely linked.

Temperate
A broad area between more extreme tropical and Arctic climates, without marked extremes of temperature or rainfall.

Thermals
"Bubbles" of rising air, produced as areas of bare or light-coloured ground warm the air above them in strong sunshine; never produced over water.

Thorax
The part of the body between the head and the abdomen.

Topsoil
Uppermost layer of soil in which decaying leaves decompose and from which roots of plants and fungi extract nutrients.

Transpiration
The loss of water vapour from leaves of plants.

Tropical
Area between the Tropic of Cancer and Tropic of Capricorn, extending across the equator, typified by high temperatures, lack of marked seasonality, and little change in length of days.

Understorey
Shrub and sapling layer in forest or woodland that is below mature trees, but above the herbaceous layer.

Veil (of fungi)
Partial or universal veil encloses growing cap and stem; splits to leave remnant ring on stipe.

Volva
A bag or cup-shaped structure at the base of the stem of a fungus, a remnant of the veil.

Index

A

access, rights of 40
acorns 96, 101
adders 70, 120
adonis blue butterflies 136
alders 87, 153, 164
algae 92, 111, 206
alpine grasslands 133
alpine plants 146-7
alpine sow-thistle 146
ammonites 152, 190, 201
amphibians 12
  heathland 126
anemometers 28
anemones 176, 180, 184, 185
antler fungus 112
antlers 116, 153
ants 62, 109, 122, 137, 201
Apollo butterflies 133, 152
arable land 68
Arctic 216-7
arrow grass 205
ascalaphids 129
astronomy 32-3
auks, little 217
avocets 187

B

bacteria 92
badgers 55, 90, 103, 104, 105
bark 94-5
bark beetles 94
barnacles 181, 190
barns 69
barometers 29
bass, European 204
bats 48, 50
  in caves 154
  in gardens 55
  nest boxes 61
beaches 176-7, 190-1
bearded vultures 151
bears 96
  polar bear 31, 212, 216
  tracks 104
beautiful yellow underwing moths 124
beavers 158, 164
bedstraw, yellow 139
bee orchids 134

beech martens 105, 143
beech trees 82, 84, 85, 87, 89, 95
bees 61, 122, 144
beetles 74-5, 92
  downland 134
  forests 90, 94
  grasslands 136
  heathland 127
  ponds 171
  scrubland 122
belugas 216
Berlese funnels 74
berries 96
betony, wood 44, 78
bilberries 108, 112
binoculars 37
biome 10
birch trees 87, 89, 95, 101, 126
birds 12
  bird baths 60
  birds of prey 150-1
  calls and songs 98
  cliff colonies 198-9
  farmland 77
  feathers 79, 150
  feeding 58-9
  flight 57, 150, 166
  flocking 72-3
  forests 98-9, 103
  freshwater habitats 160, 164
  garden birds 56-9
  migration 15, 99
  mudflats 206
  nests 51, 56, 61, 79, 99
  reedbeds 172
  rivers and lakes 158, 159
  salt-marsh 205
  scrublands 122
  shorebirds 186-7
  in trees 94
  water birds 166-7
bitterns 172
blackbirds 58, 98
blackberries 79, 84
black cherries 96
blackthorn 124-5
blue butterflies 16, 126, 135, 136, 201
  life cycle 137
bluebells 30-1, 91, 100
boar, wild 102, 104
bog asphodel 126, 173
bog bean 126, 173
bogs 159, 172-3, 212

bootlace worms 207
boundaries 69, 70-1
bracken 85
bracket fungus 94
breezes 26
brittlehead, tufted 112
brown argus butterflies 71
bryony, black 101
buckthorn, sea 178
buddleia 65
bugle 100
bulrush 161
bumblebees 178
burdock 139
burnet moths 196, 201
butterflies:
  conservation 16
  forests 84, 86, 103
  gardens 64
  grasslands 132, 133 134-5, 136, 138
  heathland 126
  scrubland 128
butterfly orchids 146
buzzard feather 108

C

cabbage, wild 197
caddis flies 160, 170
cameras 42-3, 58
camouflage 214
canopy, trees 86, 96-7
capercaillie 106
caprid, mountain 145, 149
carbon dioxide 30, 88
carnivores 14
carp 159
caterpillars 64, 97, 102, 122, 179
caves 154-5, 195
centaury 134
centipedes 62, 92, 100
chaffinches 99
chalk cliffs 194
chamois 145, 148
chamomile 138
cherry trees 87, 89, 95, 102
chestnuts 78, 87, 95, 101
chiffchaffs 99
chlorophyll 88, 89
choughs, alpine 143
cicadas 122
cinnabar moths 134
circadian rhythm 54
cirrus clouds 22, 23, 28
cladonia lichen 201
cleopatra butterflies 128
cliffs 182, 183, 194-201

climate 20-1
  see also weather
climate change 30-1, 217
clothes moths 50
clothing 38-9
clouds 22-3, 24, 28
coasts 174-209
  beachcombing 190-1
  beaches 176-9
  cliffs 194-201
  rockpooling 184-5
  sand and shingle 182-3
  sandy beaches 192-3
  seal colonies 188-9
  shorebirds 186-7
  tides 180-1
  wetlands 202-7
cockchafers 74
cockles 206
cocoons, moth 124
cold fronts 28
colonies 188-9, 198-9
Colorado beetles 74
colour
  fungi 111
  leaves 89
colt's foot 201
columbine 87
comets 33
comma butterflies 31
compasses 40
compost 62
condensation 22, 25
conifers: bark 95
  cones 113, 114-5
  forests 106-11, 112-3
  montane 106
  needles 88, 152
conifer tuft fungi 112
conservation 16
constellations, stars 32
contrails 23
coots 159
copper butterflies 138
coral 177
coral snakes 13
Coriolis effect 21, 26
cork oaks 128
cormorants 163
cornflowers 68
couch grass 139
Countryside Code 40
courtship, grebes 167
cowberries 107
cowpats 77
crabs 181, 185, 191, 193
crab apple blossoms 78
crab russula mushrooms 124
cranes 132, 133
crepuscular animals 14, 54

crickets 122, 134, 136, 138, 169, 205
crossbills 109, 115
crows, carrion 57
crustaceans 168
cuckoo spit 196
cumulus clouds 22, 23, 29
curlews 187
currents, ocean 20-1, 209
cuttlefish 191
cyclones 26
cypresses 172
cytisus 124

D

daisies 48, 132, 138
damselflies 161, 162, 164, 170
dawn chorus 54, 98
deciduous trees 82-5, 100-3
deer: antlers 153
  bones 112, 125
  fawns 105
  rutting 116-7
  tracks 104
demoiselles 162
deserts 11, 21
detector, bat 36
dew 25
dippers 158, 162
ditches 70
diurnal animals 14, 54, 64
diversity of life 12-13
dolphins 198, 208
dor beetles 75,135
dove, collared 48
downland 132, 134-5
dragonflies: on beaches 178
  eggs 171
  on heathland 126
  in lakes 159, 160, 161
  nymphs 164, 171
  on riverbanks 164
ducks 159, 160, 166-7
dunes 183, 192-3
dung 79
  beetles 77
  flies 77

E

eagles 145, 151
eagle owls 148
Earth, climate and seasons 20-1

earthworms 62, 87, 90, 91
earwigs 62, 103
echolocation 36, 44, 55
echo meter 36
edelweiss 142
egrets 205
Egyptian vultures 151
elephants 12
elm trees 89
emperor moths 122
equipment 34–45
erosion 176, 177, 182–3, 194–5
estuaries 163, 202, 206
European bison 16
European pond turtle 159
evergreens 30, 108–11
evolution 12–13

F

falcons 151
fallow deer 104, 105
false death cap mushrooms 124
fan-worms 206
farmland 66–79
feathers 79, 150
feeding birds 58–9
fens 159, 172–3
ferns 92, 93, 158, 196
field notes 44
field scabious 125
fields 66–79
finches 57
fires 120, 121, 132, 133
fish 12
  freshwater habitats 159, 160, 163
  shoals 209
fleas 50, 168, 169
flight, birds 57, 150, 166
flocking birds 72–3
floss silk trees 95
flowers 13
  downland 132
  mountain plants 146–7
  pollination 71, 96
  woodland 91, 102
fly agaric 101
flycatchers, pied 98
flying squirrel 117
fog 25
folklore, weather 29
footprints see tracks
forecasting weather 28–9

forests and woodlands 11, 80–119
  animals in 104–5
  birds 98–9
  coniferous 106–11, 112–3
  deciduous 82–5, 100–1
  forest floor 90–1
  the forest year 102–3
  fungi 110–1
  heaths and 120
  pinewoods 108–9
  recycling in 92–3
  trees 86–7
forget-me-nots 100
fossils 190, 201
foliose lichen 113
foxes 14, 54, 55
  Arctic 212, 214–5
  cubs 102
  earths 105
  eyes 55
  tracks 104
foxgloves 83
freshwater habitats 156–73
  birds 166–7
  lakes 160–1
  pond dipping 170–1
  riverbanks 164–5
  rivers 162–3
  surface dwellers 168–9
  swamps, bogs and fens 172–3
  types of 158–9
fritillary butterflies 83, 112, 134
frogs 101
  in caves 154
  heathland 126
  ponds 170
  swamps 159
  tadpoles 171
  wetlands 173
frost 25
fruit 96
fungi 13
  forests 101, 110–112
  on rotting wood 93
  scrubland 124

G

galls 97
gall wasps 97
gannets 198–9, 209
gardens 48, 56–65
  birds 56–9
  butterflies and moths 64–5
  wildlife gardens 60–3
garland flower 147

garrigue 121, 128–9
geckos 50
geese 166
germination 71
ghost crabs 193
glaciers 31, 144, 217
glasswort 206
glow-worms 55, 136
goats, mountain 148
gobies 185
goldfinches 57, 59
golden eagles 145
golden samphire 197
gorse 126, 127
granite 152, 195
grass snakes 163, 173
grasses, heaths 120
  grasslands 139
grasshoppers 122, 136–7, 138
grasslands 10, 130–9
  downland 134–5
  insects and invertebrates 136–7
  types of 132–3
graylings 162
great diving beetles 171
great green bush crickets 138
great water boatmen 168–9, 170
greater flamingoes 203
grebes 160, 166–7
greenfinches 57
green hairstreaks 124
grey squirrels 49, 86
ground beetles 74–5, 107, 204
guillemots 198, 199
gulls 160, 186, 191, 202, 206
gypsy moths 112

H

hail 25, 29
hair grass 139
hairybroom 125
halos, around moon 28, 29
harbour porpoises 208
harebells 13, 125
hares 76–7, 132, 142, 212, 214
Hart's tongue fern 85
harvestmen 136, 138
hawker dragonflies 126, 160, 178
hawfinches 98
hawkweeds 84
hay meadows 69
hazel trees 101, 102
heather 113, 120, 126–7

heaths and scrublands 118–27
  animals 122–3
  garrigue 128–9
  heathland 122, 126–7
  plants 124–5
  types of 120–1
hedgehogs 14, 61, 104, 105
hedges 69, 70–1
helleborines, red 146, 153
herb, Robert 155
herbivores 14
hermit butterflies 152
hermit crabs 185
herons 163, 164, 165
hibernation 15, 103
hides 43, 58
hillsides see mountains and hillsides
holly 87, 89, 92, 95
honey fungus 101
hornbeam 87, 89
horned poppies 178
horseflies 152
houseflies 50
houses 48, 50–1
Howard, Luke 22
hummingbirds 13
humpback whales 15
hurricanes 26

I

ibex 148
ice 211, 216–7
icebergs 217
insects 13
  in bark 94
  catching 37, 91
  on forest floor 90
  freshwater habitats 168–9
  in grasslands 132, 136–7
  insectivorous plants 172
  pollination by 96
  on scrublands 122–9
invertebrates 136–7, 207
irises, yellow 173
italian wall lizard 143

J

jackdaws 98
jasmine, rock 147
jays 57

jellyfish 181, 190–1
jersey tiger moths 65
jet streams 20
juniper 113, 153

K

kestrels 71, 54
Kentish glory moths 112
kidney vetches 147, 201
kingfishers 160, 164
kittiwakes 198
krill 216

L

ladybirds 14–15, 60, 74, 79, 109
lady's-tresses orchids 108
lakes 157, 159
  lakeshore walk 160–1
  surface dwellers 168–9
lapwings 77
large blue butterflies 137
larvae 91, 92, 93, 94
laurel, mountain 146
lavender 129, 132, 153
leaf litter 90
leaf miners 97
leaves 88–9
lemmings 212, 214
lenticular clouds 22
lesser celandine 91, 100
lichens 83, 92, 111, 113, 152
life, diversity of 12–13
lightning 27
limestone 154, 194, 201
limpets 185, 190
ling 108, 109
lizards 14, 127
  basking 71
  garrigue 121
  heathland 127
  mountains 143
  wetlands 173
local habitats 48–9
logs 61, 92–3
long-legged flies 169
longhorn beetles 74, 92
lugworms 181, 206
lyme grass 192
lynx
  Eurasian 148
  European 117

M

madwort, mountain 71
magpies 62, 132
maidenhair ferns 196
maize 78
malachite beetles 136
mallard 159, 167
mallows 200, 205
mammals 12
  mountains 148-9
  riverbanks 164-5
  shelters 105
  tracks 104
  wildlife gardens 61
mandarin ducks 167
map butterflies 79
maples 89, 95
maple wings, red 101
maps 28, 40
maquis 121
marmots 133, 144
marram grass 193
marsh fritillary 83
marshes, salt 202, 204-5
mayflies 158, 160
mayweed, sea 197
creeping mazus 147
meadows 69, 144, 146
mermaids' purses 190
metamorphosis 64
meteorology 28-9
mice 50, 51, 69
microclimates 26, 144
migration 15, 99
mimicry 13
minks 70, 164
minke whales 208
mist 25
mistletoe 103
mistrals 26
moeritherium 12
moles 68, 76, 79
moltkia, alpine 147
money spiders 52
monsoon 26
montpelier snakes 203
Moon 32
  halos around 28, 29
  and tides 180-1
moorhens 70
moorland 120
morels 101, 106, 112
mosquitoes 60, 213
mosses 85, 93, 108, 158-9
moths: attracting 55
  in caves 154
  cocoons 124
  in coniferous forests 112
  in deciduous woodlands 82, 100

moths (cont.)
  in downland 134
  forests 94, 103
  gardens 64, 65
  in grassland 138
  in houses 50
  scrubland 122
  trap 36
mountains and hillsides 140-55
  birds of prey 150-1
  caves 154-5
  mammals 148-9
  plants 146-7
  zones 144-5
mountain hares 142
mudflats 203, 206-7
mudshrimp 207
mushrooms 68, 109, 110-1, 124
mussels 176, 184, 190, 206
myrtle 120, 128

N

narwhals 208
National Phenology Network 31
Nature's Calendar 31
navigation, migration 15
needles, conifers 88, 152
nests: birds 51, 56, 61, 79, 99
  mice 51
  wasps 51
newts 126, 145, 165, 171
nightjar 120
nimbus clouds 22
noctilucent clouds 22
nocturnal animals 14, 54, 55, 105
notebooks 44
nuthatches 59
nuts 78, 96, 101

O

oak beauty moths 82
oak trees 84, 89
  cork oaks 128
  English oaks 95
  galls 97, 101
  sessile oaks 87
observation 36
ocellated lizards 121
oceans 208-9
  currents 20-1, 209
  sea level rises 31
  see also coasts
olives 79, 124

olive trees 121
olms 155
onagers 133
orange peel fungus 101
orb spiders 52-3, 152
orcas 208-9
orchids: downland 132, 134
  farm and field 68
  grassland 132, 139
  mountains 146
  parasitic 91
  scrubland 129
  woods 108
ospreys 16
otters 16-17, 164, 165
owls
  barn 69
  pellets 79, 125
  snowy 14, 213, 214
  tawny 55, 82
oxygen 88, 145
oystercatchers 187
oysters 190

P

pandas, giant 13
parkland 83
parks, urban 49
parsley, hedge 78
partridges, grey 77
pasque flower 132
pasture 68
peacock butterflies 65, 79
peacock worms 207
peaks, mountains 143
peat 172
penguins 12
penny bun mushrooms 124
peregrine falcons 72-3, 151
periwinkles 180
permafrost 212
pests 50
phalaropes 213
pheasants 43, 77
pheromones 122
photography 42-3, 58
photosynthesis 88
pine cones 29, 113, 114-5, 152
pine hawk moths 112
pine martens 117
pine trees
  forests 107, 108-9
  garrigue 121
  mountains 144
  plantations 106
pinnacles, mountains 142

pitcher plants 172
plane trees 95
planets 33
plantains, buck's horn 200
plants
  in cave entrances 155
  on cliffs 196-7, 200-1
  and climate change 30
  flowering 13
  following the Sun 54
  freshwater habitats 158
  fungi and 111
  garrigue 128-9
  grasslands 138-9
  heaths and scrublands 120-1, 124-5
  insectivorous 172
  leaves 88-9
  mountains 146-7
  sandy beaches 192-3
  scrubland 124-5
  water plants 63
  wetlands 172-3
plaster casts, of tracks 104
plateaux 142
plovers 179, 186
poisonous fungi 110
polar climate 20
polar regions 216-7
pollination 96
pollution, oceans 209
pond skaters 159, 168-9
ponds 159
  gardens 63
  heathland 126
  pond dipping 170-1
pondweed, Canadian 159
pools, rock 184-5
pooters 37, 91
poplar trees 89
poppies 68, 178, 213
prawns 185
praying mantis 124
predators
  freshwater habitats 164
  mountains 148, 150-1
  oceans 209
prints see tracks
Project BudBurst 31
pseudoscorpions 154
ptarmigans 142
puffball fungi 110, 112
puffins 199
purple and orange fungus 112
purple emperor butterflies 84
purple hairstreaks 86

Q

quadrats 137
quaking grass 139
quartz 152

R

rabbits 132, 136
raft spiders 126, 168
ragworms 207
ragworts 49, 193
railways 49
rain 24, 28-9
rainbows 29
rainforests
  climate 21
  temperate rainforest 83
raptors 150
raspberries 113
rats 49, 62
rauli 89
razor-shells 190
record-keeping 36, 44-5
recycling 62, 92-3
red deer 116-7, 153
red squirrel 12
redstarts 56, 98
reeds 10, 159, 172, 173
reedmace 161
reindeer 15, 212-3
reptiles 12
  gardens 54
  heaths and scrublands 120-1, 128, 122
rest-harrow 139
rhododendrons 146
ringlet butterflies 145
rivers 157, 158, 162-3
robins 99
rock roses 147
rocket, sea 192
rockpooling 184-5
rocks: beaches 176
  cliffs 194-5
  erosion 176, 182-3
  mountains 152
rosemary 121
roseroot 200
rove beetles 74
rowan 153
rushes, flowering 160

S

safety 40-1
sage 122, 125
St John's wort 147
salamanders 90, 101, 154
salinas 203

salmon 12, 163
salt marshes 202, 204-5
saltwort, prickly 192
samphire, rock 196
sand 176, 182-3, 192-3,
sand-dollars 190
sand traps 55
sanderlings 176
sandpipers 187, 203
sandstone 195
sandwort, sea 192
sardines 209
Sardinian warbler 121
saw wort 139
saxifrages 142, 155
scallops 188
scarab beetles 74
scarce swallowtail
    butterflies 124
scarlet darter butterflies
    173
scarlet kingsnake 13
southern hawker 126
    178
scavengers 14
scorpions 122, 170
Scots pines 107, 113,
    114
scrublands see heaths
and scrublands
sea beans 191
sea beet 200
sea holly 176, 193
sea kale 177
sea lavender 200, 202
sea lions 188
sea mayweed 197
sea slaters 180
sea stars 181
sea urchins 185, 191
seals 188-9, 216
seasons 20
seaweeds 29, 180-1,
    185, 191
sedges 153, 159, 172
sediment, coasts 182
sedimentary rocks 183,
    195
seeds 71, 96, 114-5
shags 199
shearwaters 209
shelducks 167
shells 177, 190-1
shingle 177, 182-3
shorebirds 186-7
shore crabs 204
showers, rain 29
shrews 69, 90, 104,
    125, 134, 165
shrikes 122
shrubs 87
siskins 59
sketching 45
skipper butterflies 138

skuas 198
skylarks 68
slate 152
slime mould 110
slow worms 138
slugs 91
small elephant
    hawkmoths 64
small pearl-bordered
    fritillary butterflies 112
snails 76, 90, 136, 168,
    170, 201, 206
snakes 54
    adder 70
    coral 13
    discarded skin 79
    grass 163, 173
    mimicry 13
    Montpelier 203
    heathland 120, 127
    swamps 159
    swimming 163
    wetlands 173
snipe 173
snow 25, 104
soil 87, 90-1
soldier beetles 79
Solomon's seal 84
solstices 20
songbirds 54
sorrel 91, 155
Spanish festoon
    butterflies 128
species 12
spiders 52-3, 112
    caves 154
    forest floor 90
    gardens 52
    grasslands 136
    heathland 126
    in houses 50
    scrublands 122
    webs 93
sponges 190
spore prints, fungi 111
spring squills 196
springs 17
springtails 77
spruce trees 106, 107,
    113
spurges, sea 192, 193
squirrels 49, 54, 86,
    100, 105, 115
stag beetles 61, 75
stalactites and
stalagmites 154
starfish 185, 191
starlings 56, 72-3
stars 32-3
steppes 133
sticklebacks 171
stinkhorns 110
stone pines 143
stone walls 69, 71
stonecrops 177, 200

stormy weather 26-7
stratus clouds 22
strawberry trees 121,
    128
streams 157, 158, 162
streets 49
Sun 20, 54, 88
sundews 126, 172
sunflowers 54, 56
sunsets 29
superstitions, weather
    29
swallows 51
swallowtail butterflies
    64-5, 124
swallow-tailed moths
    100
swamps 159, 172-3
swans 166
sycamores 95, 101
symbiosis 111, 146

T

tadpoles 102, 171
taiga 107
teal 205
telescopes 33, 36
temperate rainforest
    83
termites 50, 138
terns 177, 186, 216
thermals 150
thermometers 29
thistles 78, 132, 139,
    153, 179, 201
thrift 201
thrushes 56, 69, 86, 90
thunderstorms 23, 27,
    29
thyme 135, 137, 193
tides 180-1, 204-5
tiger beetles 127, 138
tits 11, 57, 61, 172
toads 102, 126, 193
tongue orchids 129
tornadoes 27
tortoises 121
tracks 104
    birds 186, 206
trade winds 20, 21
traps 55
traveller's joy 78
tree echium 125
tree lungwort lichen 83
trees 13, 86-7
    bark 94-5
    canopy 86, 96-7
    growth rings 86
    leaves 88-9
    shapes 87
    see also forests and
    trefoils, bird's foot 201

tufted brittlehead 112
tufted duck 167
tundra 210-15
turtles 159, 192
typhoons 26

U

urban areas 49

V

valerian, red 179
viburnum 87
viewers, rockpooling
    184
violets, dog 100
viper's bugloss 178
volcanoes 30, 143, 147,
    195
voles 70, 76, 125, 134,
    165
volunteer schemes 31
vultures 150-1

W

waders 186, 187, 205
wagtails 158
walls, stone 70-1
walruses 188, 216
warblers 99, 122,
    126-7, 172
warm fronts 28
wasp spiders 52-3
wasps 48, 50-1, 97,
    112, 122-3, 152, 197
water: bird baths 60
    freshwater habitats
    156-73
    oceans 10, 20-1, 31,
    208-09
water birds 166-7
water boatmen 169, 170
water crickets 169
watercress 172
water measurers 169
water plants 63
water rails 172
water scorpions 170
water shrews 165
water vapour 22, 24
water voles 70, 165
waterfowl 160, 166
weather 18-31
    climate and seasons
    20-1
    clouds 22-3
    forecasting 28-9
    stormy weather 26-7
    wet weather 24-5

web cap mushrooms,
    124
webs, spiders 52, 93
weevils 74
wetlands 10, 126, 166,
    172-3, 202-7
whales 15, 20, 208, 216
wheat ears 78
whelks 181, 190, 206
whirligig beetles 169
white butterflies 65,
    138, 179
wildcats 107
wild garlic 82
wildlife gardens 60-3
willows 164
window boxes 60
winds 20, 21, 24, 26-7,
    192
winkles 190
wintergreen 113
wolf spiders 52, 85,
    112
wolverines 107, 212
wolves 117, 148, 212,
    214
wood anemones 100
woodlark 69, 120
woodlice 90
woodpeckers 57, 59,
    94, 115, 143
woodpigeons 99
woods see forests and
woodlands
woodworm 50
wormery 62
wracks 176, 180, 185

Y

yellow shell moth 136
yew 106

Z

zebra spiders 52
zinnia 49

Acknowledgments

Dorling Kindersley would like to thank the following for their help in the preparation of this book: George McGavin for consultancy, Daniel Gilpin and Elizabeth Munsey for editorial assistance, Sunita Gahir for design assistance, Hilary Bird for indexing, Rakesh Kumar for the jacket, Shanker Prasad for CTS assistance, The Cotswold Store and Alana Ecology for supplying equipment, and Peter Anderson and Gary Ombler for additional photography. The publisher would also like to thank the following for their kind permission to reproduce their photographs:

(Key: a-above; b-below/bottom; c-centre; f-far; l-left; r-right; t-top)

1 Corbis: Visuals Unlimited (cr). Dorling Kindersley: Frank Greenaway courtesy of Natural History Museum, London (cl). 4 Dreamstime.com: Cosmin Manci (cr). 5 Dorling Kindersley: Frank Greenaway courtesy of Natural History Museum, London (tc); Stephen Oliver (b). 6 Dorling Kindersley: Neil Fletcher (bc) (fbl); Matthew Ward (br); Jerry Young (cl). 7 Alamy Stock Photo: Bob Gibbons (r). Dorling Kindersley: Frank Greenaway courtesy of Natural History Museum, London (br); Derek Hall (bl) (bc/r). 8 Corbis: Gisuke Hagiwara / amanaimages. 10 Alamy Stock Photo: Paul Glendell (bl). Corbis: Patrik Engquist (br); Frank Krahmer (c); Roger Tidman (r). 11 Ardea: Francois Gohier (cr/prairie dogs) (cra) (tr). Corbis: W. Cody (ca); Simon Weller (bl). Dorling Kindersley: NASA / Finley Holiday Films (tr). Getty Images: Robert Postma (crb). Photolibrary: Shattil and Rozinski (cr/ferret). 12 123RF.com: (c); taviphoto (tr). Dorling Kindersley: Getty RF: morgan stephenson (t). 12-13 Alamy Stock Photo: WILDLIFE GmbH. 13 123RF.com: anest (tl). Alamy Stock Photo: Rolf Nussbaumer (bl). Corbis: amanaimages (c); Scott Stulberg (cb); Herbert Zettl (bc). Dorling Kindersley: Dreamstime.com: Jason Ondreicka / Ondreicka (c); Harry Taylor, courtesy of Natural History Museum, London (b). Dreamstime.com: Prambuwesas (tc); Dr Sauer (c). 14 Dreamstime.com: Helen Davies (b). Alamy Stock Photo: All Canada Photos (b). 14-15 naturepl.com: Jane Burton. 15 Getty Images: Jeff Hunter (tr); Joel Sartore / National Geographic (tr). Dreamstime.com: Aleksander Bolbot (tr). Corbis: E & P Bauer (br). Getty Images: Christopher Furlong (c). 16-17 Alamy Stock Photo: Peter Arnold, Inc. 18 Corbis: Visuals Unlimited (br). 20 Corbis: Paul Souders (tr); Hubert Stadler (br). Getty Images: Astromujoff (br). 21 Corbis: Theo Allofs (bl). Getty Images: Philip and Karen Smith (tr). 22 Corbis: Mike Theiss / Ultimate Chase (crb). John Marshall (tr). 23 Getty Images: Mike Hedges (bc); blickwinkel (bc); Ashley Cooper (ca); Eddie Gerald (clb); NaturePics (c); Ken Walsh (bl). 24 Corbis: Johnathan Smith; Cordaiy Photo Library Johnathan Smith; Cordaiy Photo Library Ltd (cb); Craig Lovell (cla). 24 Alamy Stock Photo: Andy Arthur (tr); Ryan McGinnis (crb). Dreamstime.com: (cra) Getty Images: Jamey Stillings (br). 25 Alamy Stock Photo: Ashley Cooper (crb); imagebroker (cl); Oleksiy Maksymenko (bl). Corbis: Glowimages (c); Richard T. Nowitz (br). Getty Images: Arctic Images (c); Jeff Foott (cl). Science Photo Library: Kenneth Libbrecht (tr) (cra) (tcra). 26 Alamy Stock Photo: Hemis (c). 26-27 Corbis: Peter Wilson. 27 Corbis: Gene Blevins (bc). Getty Images: Alan R Moller (tr); Priit Vesilind / National Geographic (bl). Science Photo Library: J. G. Golden (ca). 28 Dreamstime.com: Roman Ivaschenko (cl). Alamy Stock Photo: WildPictures (cl). Getty Images: Brian Stablyk (cb); Kim Steele (cl). iStockphoto.com: (br). NOAA: Carol Baldwin (bc). Image courtesy of Oregon Scientific (UK) Limited: (tc/bottom) (tc/middle). 29 Alamy Stock Photo: Nepal Images (tc/top). Corbis: Jorma Jamsen (br); Tony Hallas / Science Faction (crb). Getty Images: Michael McQueen (c). Science Photo Library: Garry D. McMichael (c). 30 Corbis: Larry Dale Gordon (cra). 30-31 Alamy Stock Photo: Jeremy Walker (cro). 31 Corbis: Laura Sivell / Papilio (br); Chen Zhanjie / XinHua Press (cla). Getty Images: Johnny Johnson (ca); Visuals Unlimited (crb). 32 Getty Images: Lew Robertson (bra). 32-33 Till Credner & Sven Kohle / allthesky.com. 33 Alamy Stock Photo: Peter Arnold, Inc (cr); Galaxy Picture Library (cra). Corbis: Roger Ressmeyer / Science Faction (cl). 36 Dreamstime.com: Mikhail Primakov (r). Magenta Electronics (br). 36 SWAROVSKI OPTIK: (tr). Watkins & Doncaster: Wildlife Acoustics, Inc. 36 37 FLPA: Hugh Clark (b). rspb-images.com: Malcolm Hunt (tl). 37 RICOH Imaging Europe S.A.S: (tr). 40 Dorling Kindersley: 123RF.com: Aleksey Boldin (cr). Alamy Stock Photo: Peter Titmuss (bc). 40-41 Getty Images: Superstudio. 41 Getty Images: Joel Sartore / National Geographic (c). 42 123RF.com: cloud7days (bc). Canon Inc: (cr). Dreamstime.com: Dimitry Romanchuck (c); Audrius Merfeldas (cb); Audrius Merfeldas (b). FLPA: Yva Momatiuk & John Eastcott (crb). Getty Images: Gerry Ellis (tr). Pentax UK Ltd: (br/room). Sigma Corporation: (br/wide). 42-3 Corbis: Peter Johnson (t). 42-43 Getty Images: Gary W. Carter (b). Getty Images: David Maitland (cl). 43 Corbis: Michele Westmorland (cra). FLPA: Robert Canis (b); David Hosking (cra); Roger Tidman (cb). Getty Images: Peter Lilja (tl). rspb-images.com: Niall Benvie (br); Gerald Downey (cl). 44 Dreamstime.com: Fedsax (tl). Aleksey Boldin (br). Dorling Kindersley: Fedsax (tl). Wildlife Acoustics, Inc.: (ca). Marek Walisiewicz (c). 45 Corbis: Bob Krist (cr). Marek Walisiewicz (c). 48 123RF.com: sandermeertinsphotography (c). Alamy Stock Photo: blickwinkel (c). Corbis: Elizabeth Whiting & Associates (tl). Getty Images: Evan Sklar / Botanica (br). 49 123RF.com: lightpoet (cb). Corbis: Toshi Sasaki / amanaimages (tl); Owaki-Kulla (bl). Getty Images: Mitchell Funk (c). 50 123RF.com: vitalisg (fcl). Alamy Stock Photo: Carl Corbidge (cla).

FLPA: Erica Olsen (cla). naturepl.com: John Downer (bc); Kim Taylor (crb). Photolibrary: Roger Jackman (bl). Science Photo Library: Tom McHugh (c). 51 Getty Images: Tony Bomford / Photolibrary (br). NHPA / Photoshot: Stephen Dalton (br). 52-53 Alamy Stock Photo: imageBROKER. 54 Corbis: George McCarthy (bc). Getty Images: Oxford Scientific / Photolibrary (c). naturepl.com: Laurent Geslin (tl). 55 Corbis: O. Alamany & E. Vicens. FLPA: Malcolm Schuyl (br). Getty Images: Raymond Blythe / Photolibrary J & C Sohns (crb). 56 123RF.com: chepko (bl); cosmin (br). 64-65 Alamy Stock Photo: Stan Kujawa. 65 123RF.com: geki (tc). Getty Images: Yann Layma (bc); George Grall / National Geographic (crb). 68 Corbis: Erenie Janes (br). 69 Dorling Kindersley: Kim Taylor (c); Rollin Verlinde (cl). 70 Alamy Stock Photo: Itsik Marom (br). Nigel Cattlin (cl). Corbis: Michael Rose, FLPA (bl); Joe McDonald (c); Roger Tidman (cb). Getty Images: David Zimmerman (tr). iStockphoto.com: (c). 71 123RF.com: mortenekstroem (cb); sandermeertinsphotography (b). Dreamstime.com: Michael Smith (cl). Alamy Stock Photo: imagebroker (clb). Getty Images: Mad MaT (crb); Visuals Unlimited (clb). 72 naturepl.com: Bruno D'Amicis (bl) (br). 72-73 Alamy Stock Photo: MediaWorldImages. 74 123RF.com: cosmIn (clb). Alamy Stock Photo: Neil Hardwick (c). Getty Images: Brian Hagiwara (c); GK Hart / Vikki Hart (c); Oxford Scientific / Photolibrary (cra). Science Photo Library: Bjorn Svensson (tr). 74-75 Alamy Stock Photo: flabCC. 75 123RF.com: creativenature (clb). Corbis: Tom Bean (cla). FLPA: Nigel Cattlin (bc). Getty Images: Michael Leach / Photolibrary (c). Science Photo Library: Martyn F. Chillmaid (br); Terry Mead (cla). 76-77 naturepl.com: Gary K. Smith. 77 123RF.com: fotobird328 (cr). Alamy Stock Photo: Nigel Cattlin (cr/springtail); Tom Joslyn (crb/dung fly). Corbis: PULSE (br). Getty Images: Andrew Howe (c); Roger Jackman / Photolibrary (bl); Ann & Steve Toon (tr). 78-79 Alamy Stock Photo: flabCC. 79 Alamy Stock Photo: Buiten-Beeld (br). 82 Alamy Stock Photo: JTB Photo Communications, Inc (br). Dorling Kindersley: Frank Greenaway courtesy of Natural History Museum, London (br); Sean Hunter (tr). Getty Images: David Tipling (tl). 83 123RF.com: atosf (bl). Dreamstime.com: Digitalimagined (br). Alamy Stock Photo: David Noble Photography (tl). Corbis: Larry Lee Photography (bl). naturepl.com: SCOTLAND: The Big Picture (br). 84 Alamy Stock Photo: blickwinkel (ca). 85 Alamy Stock Photo: Premaphotos (cla). Dreamstime.com: Chuanthit Kunlayanamitre (tl). naturepl.com: Will Watson (b). 86 123RF.com: creativenature (tl). Alamy Stock Photo: Nature Photographers Ltd (br). Dorling Kindersley: Frank Greenaway courtesy of Natural History Museum, London (tr). Getty Images: Photolink (br); Visuals Unlimited (c). 88 Getty Images: Jozsef Szentpeteri / National Geographic (cla). 89 Dreamstime.com: Emanoo (cra). Corbis: Karl Kinne (cra). Rob Herr: (br). 90 123RF.com: o2beat (c). Alamy Stock Photo: Simon Colmer and Abby Rex (tr). Dorling Kindersley: Rollin Verlinde (cr). 91 Dreamstime.com: Rbiedermann (cr). Alamy Stock Photo: David Crausby (tl). Getty Images: Travel Ink (tr). 92 123RF.com: marcouliana (bl). 94 123RF.com: gucio_55 (cla). Alamy Stock Photo: Peter Arnold, Inc (cra); Mark Breturton (br); John Glover (br); Krystyna Szulecka Photography (bc). Corbis: Roger Tidman (t). naturepl.com: Jussi Murtosaari (cr). 94-95 Alamy Stock Photo: Ashley Cooper. 95 Alamy Stock Photo: blickwinkel (bl); WILDLIFE GmbH (cra). 96 Alamy Stock Photo: blickwinkel (br); INTERFOTO (cb). naturepl.com: Staffan Widstrand (bl). 96-97 Getty Images / iStock: Christopher OHara. 97 Alamy Stock Photo: blickwinkel (bl); David Chapman (ca). 98 Dreamstime.com: Aniszewski (br); Juan Carlos Martinez Salvadores (bl). naturepl.com: Mark Hamblin (c). 99 Alamy Stock Photo: Ian Rutherford (cla). Dreamstime.com: Gabe9000c (c). Dorling Kindersley: Peter Chadwick, courtesy of the Natural History Museum, London (c). 100 Dreamstime.com: Caymia (br). Rbiedermann (cra). 102 Alamy Stock Photo: Michael Griffin (bc). Corbis: Lothar Lenz (tr); Manfred Mehlia (cra). Getty Images: Stephen Studd (cr). naturepl.com: Philippe Clement (bl). 103 Dreamstime.com: Jaroslav Frank (cra); Anna Kravchuk (clb). Corbis: David Chapman Natural Selection (cl); Fridmar Damm (cra). Getty Images: Visuals Unlimited (cr). 104 Alamy Stock Photo: David Hosking (tr). Corbis: Niall Benvie (bl). Dreamstime.com: Daniel Rodriguez Garriga (br). 104-105 Getty Images: Joel Sartore / National Geographic. 105 Alamy Stock Photo: Papilio (bl). Dreamstime.com: Vladimir Cech (tl). naturepl.com: Paul Johnson (tr); Colin Seddon (br). Getty Images: Charcrit Boonsom (tl). 107 Alamy Stock Photo: Picture Scotland (tr). Corbis: W. Perry Conway (cb); Steve Austin; Papilio (ca); Ed Darack / Science Faction (br). 108-109 Alamy Stock Photo: Louise A Heusinkveld. 110 Alamy Stock Photo: A. P (tr); Neil Hardwick (l). 112 Dreamstime.com: Jamesanahlon (cb); Jens Stolt (bl); Whiskybottle (tr). 114 Getty Images: Peter Lilja (c). 114-115 Alamy Stock Photo: First Light. 115 Alamy Stock Photo: Arco Images GmbH (tr); Juniors Bildarchiv (tr). Getty

Images: Bob Stefko (cra). 116-117 Getty Images: Raymond K. Gehman. 117 Corbis: Steven Kazlowski / Science Faction (tr). Dreamstime.com: Ivanka Blazkova (cra). Getty Images: Riccardo Savi (crb); Ronald Wittek (br). 118 Dorling Kindersley: Frank Greenaway courtesy of Natural History Museum, London. 120 Corbis: Julie Meech; Ecoscene (tl); Pam Gardner; FLPA (cr). Getty Images: Adam Burton (br). 121 Alamy Stock Photo: Marcos G. Meider (clb); David Boag (cr). Dreamstime.com: Office2005 (cla); Steprphotos (bl); Vaeenma (br). 122 123RF.com: suerob (tr). Alamy Stock Photo: Marcus Gosling (cla). Dorling Kindersley: John Keates, courtesy of Natural History Museum, London (c). Dreamstime.com: Sandra Standbridge (tl). FLPA: Mark Winwood / Photolibrary (br). Getty Images: Manfred Pfefferle / Photolibrary (cb). Shutterstock.com: Macronatura.es (clb). 123 Alamy Stock Photo: blickwinkel. 124 Alamy Stock Photo: Nature Photographers Ltd (c). Dreamstime.com: Agami Photo Agency (cla); Cosmin Manci (cr). 124-125 Alamy Stock Photo: Hoberman Collection UK (c). 125 Dorling Kindersley: Neil Fletcher (tc); Frank Greenaway courtesy of Natural History Museum, London (cla). Still Pictures: WILDLIFE / N Benvie (crb) (br). 126 Alamy Stock Photo: Nigel Pye (clb). Corbis: doc-stock (cra); Sally A. Morgan, Ecoscene (tr). naturepl.com: Jane Burton (ca). 126-127 Alamy Stock Photo: Hoberman Collection UK. 127 Alamy Stock Photo: Arco Images GmbH (bc); Juniors Bildarchiv (br). Getty Images: Paulo De Oliveira / Photolibrary (c). 128-129 Alamy Stock Photo: David Boag. 129 Corbis: 132 Corbis: Frank Blackburn, Ecoscene (tl). Dorling Kindersley: Sean Hunter (c). Dreamstime.com: Lianem (br); Svitlana Tkach (bl); Sweetsake (bc). 133 Alamy Stock Photo: Peter Arnold, Inc (cr); WILDLIFE GmbH (clb). Corbis: Eric and David Hosking (bc) Dreamstime.com: Alanjeffery (bl); Tommaso Barbanti (br); David Havel (c); Isselee (cra). 134 Corbis: George McCarthy (cb). Dreamstime.com: Jamesanahlon (cb); DEA / V. Giannella (tc). 134-135 naturepl.com: William Osborn. 135 Alamy Stock Photo: Redmond Durrell (cr); Roger Wilmshurst; FLPA. Dreamstime.com: PeterWaters (cr). 136 Corbis: Hans Pfleftschunger / Science Faction (br). Dreamstime.com: Alslutsky (c). Getty Images: Raymond Blythe / Photolibrary (cl). naturepl.com: Meul / ARCO (cra). 136-137 Alamy Stock Photo: Kim Taylor. 137 Dreamstime.com: Ian Redding (fcla). T. Komatsu / Japan. Sci Rep 6, 36364 (2016): (ca). Marcin Sielezniew: (ca). 138 Alamy Stock Photo: Nature Photographers Ltd (c). Dreamstime.com: Alslutsky (cb); Darius Baužys (cla); Ovydyborets (bl). Getty Images: Jody Dole (c). 142 Alamy Stock Photo: blickwinkel (bc). Corbis: Pablo Corral Vega (bl); Tim Zurowski (cra). Dorling Kindersley: Dreamstime.com: Scattoselvaggio (br). Dreamstime.com: Lucagal (bl); Ivan Kluciar (bc). 143 Dreamstime.com: Bernard Bialorucki (tr); Gherzak (tc); Ondřej Prosický (ca); T. Van Urk (bl); Wkruck (bc); Marcelkudla (cr). 144 Alamy Stock Photo: Michael Doolittle (bl); WILDLIFE GmbH (br). Corbis: Roy Hsu (tr). Dreamstime.com: Enrico Morando (tr). Getty Images: Konrad Wothe (cr). 144-145 Alamy Stock Photo: Art Kowalsky (tr). 145 Alamy Stock Photo: Andrew Darrington (cr). Dreamstime.com: Digistockpix (bl); Pavel V (tl). Getty Images: (bl/cycling); DEA / A. Calegari (br); George F. Mobley / National Geographic (tr). 146 Corbis: David Muench (cr). Dreamstime.com: Pawel Gubernat (cr). Getty Images: Visuals Unlimited (c). naturepl.com: Kirkendall-Spring (br). 147 Alamy Stock Photo: CuboImages srl (br). 148-149 naturepl.com: Orsolya Haarberg. 148 Dorling Kindersley: Alvinge (bl). 150-151 Photolibrary: John Cancalosi. 151 Alamy Stock Photo: Buiten-Beeld (c). Corbis: Roger Tidman (cr). Getty Images: Jonathan Gale (cb); Joseph Van Os (br). Dreamstime.com: Caglar Gungor (cr). 152 Getty Images: Visuals Unlimited (ca). 152-153 Alamy Stock Photo: Bob Gibbons. 153 Dreamstime.com: Taviphoto (b). 154 Dorling Kindersley: Frank Greenaway courtesy of Natural History Museum, London (bc); Rollin Verlinde (br). Dreamstime.com: DEA / A. Calegari (cr); David Littschwager / National Geographic (bl). 154-155 Photolibrary: Luis Javier Sandoval. 155 Dreamstime.com: Pedro Antonio Salaverria Calahorra (tr); Getty Images / iStock: IvanaOK (cla). 158 Corbis: Adam Woolfitt (tr). Dorling Kindersley: Frank Greenaway courtesy of Natural History Museum, London (br); Rollin Verlinde. Dreamstime.com: Ondřej Prosický (cra). 159 Dreamstime.com: Theripper (br). Getty Images: Matt Cardy (tl); stockbyte (cr). 160-161 Getty Images: Tyler Gray. 161 Dreamstime.com: Alslutsky (cr). 162 Corbis: Niall Benvie (c). 162-163 Dreamstime.com: Luca Lorenzelli. 163 Dorling Kindersley: Kim Taylor (br); Daniel Cox / Photolibrary (cra). Getty Images: Manfred Pfefferle / Photolibrary (cra). 164 Corbis: Kennan Ward (cb). Dreamstime.com: Mircea Bezergheanu (br). Getty Images: Hauke Dressler (cl); Visuals Unlimited (br); Dominic Harcourt Webster (tl). 164-165 naturepl.com: John Cancalosi. 165 Dorling Kindersley: Cyril Laubscher (crb); Jan Van Der Voort (cra). Getty Images: Elliott Neep (tr); David Boag / Photolibrary (br). 166 Corbis: Roger Tidman (cr). Getty Images: Mikelane45 (cra). 167 Dreamstime.com: Guido Vrola (tr). Getty Images: Frank Krahmer (c); Oxford Scientific / Photolibrary (br); Gerhard Schulz (ca). 168 Alamy Stock Photo: blickwinkel (cl). Getty Images: Spike Walker (c). 168-169 Alamy Stock Photo: Papilio. 169 Alamy Stock Photo: Daniel Borzynski (br); Premaphotos (br). Getty Images: Oxford Scientific / Photolibrary (cra). 170 Alamy Stock Photo: Brian Bevan (tr). Corbis: Colin Milkins / Photolibrary (cr); Visuals Unlimited (cr). 171 naturepl.com: Fabio Liverani (tl). 172 Alamy Stock Photo: blickwinkel (bl); John Downer (bl); Frank Krahmer (cr); Sven Zacek / Photolibrary (cra); Paul E Tessier (tc). 167 Dreamstime.com: Chris Pancewicz (br). 173 Corbis: George McCarthy (ca). Dreamstime.com: Agami Photo Agency (bl); Andrew Astbury

(cb). Getty Images: Jan Tove Johansson (bl); Harold Taylor / Photolibrary (cla); Niall Benvie / Photolibrary (clb); Tohoku Colour Agency (tl); Visuals Unlimited (cr). 174 Alamy Stock Photo: Richard Murphy (tr). Corbis: Arthur Morris (bl). Getty Images: Image Source (bl); Jupiter Images (tl). 177 Alamy Stock Photo: Florapix (cla). Corbis: Michele Westmorland (cla). Chris Gibson: (bl). Dreamstime.com: Agami Photo Agency (cla). Getty Images / iStock: vuk8691 (tr). 178-179 Corbis: Andrew Brown; Ecoscene. 180 Alamy Stock Photo: John Taylor (cla); Darylyne A. Murawski / National Geographic (cla). 180-181 Corbis: Bertrand Riegel / Hemis. 181 Alamy Stock Photo: doughoughton (br); JS Callahan / tropicalpix (cra). Corbis: Eberhard Streichan (c); Anthony West (br); Adam Woolfit (cla). Getty Images: Laurance B. Aiuppy (bc). 182 Alamy Stock Photo: Leslie Garland Picture Library (tr). Dreamstime.com: Simon Taylor (cb). 183 Alamy Stock Photo: Skyscan Photolibrary (cr). Getty Images: Thierry Grun (br); Camille Moirenc (bl). 184 Alamy Stock Photo: Lightworks Media (bl); Peter Titmuss (tr). Getty Images: Bob Pool (br). 185 Dorling Kindersley: Frank Greenaway courtesy of Natural History Museum, London (cra). 186 123RF.com: Johan van Beilen (bl). Alamy Stock Photo: Robin Chittenden (c); Nick Greaves (br). 187 Dreamstime.com: Stuartan (cr). Getty Images: Rosemary Calvert (tr). 188 Alamy Stock Photo: Nigel Housden (bl). Corbis: Ronald Wittek / dpa (bc). 188-189 Corbis: Paul Darrow / Reuters. 190 Alamy Stock Photo: David Chapman (br); photow.com (cra). 190-191 Alamy Stock Photo: Stephen Oliver (t). 192 Alamy Stock Photo: Rami Aapasua (bl). Getty Images: AFP (br). naturepl.com: Juan Manuel Borrero (tr). 193 Alamy Stock Photo: DK Limited (cra). Getty Images: Frank Cezus (cra); Barrie Watts / Photolibrary (bc); Time & Life Pictures (cr). 194 Corbis: Chinch Gryzniewicz, Ecoscene (tl). Dreamstime.com: John Loader (bl); Razvanjp (br). 195 Corbis: Ashley Cooper (tr); Destinations (c); Robert Pickett (bl). Dorling Kindersley: Colin Keates, courtesy of Natural History Museum, London (br); John Loader (bl); Razvanjp (br). 197 Alamy Stock Photo: Ray Wilson (cr). 198 Alamy Stock Photo: Stefan Huwiler (tl). charliephillips.co.uk: (cr). Corbis: Joe McDonald (br). Getty Images: Ghislain & Marie David de Lossy (cla). 198 Alamy Stock Photo: Jerome Murray - CC (cla). 198-199 Getty Images: Christina Bollen / Photolibrary. 199 Alamy Stock Photo: KEVIN ELSBY (bl). Corbis: Winfried Wisniewski (br). Getty Images: Siqui Sanchez (bl). naturepl.com: Chris Gomersall (tr). 202 Alamy Stock Photo: Michael Howell (cr). Corbis: Yann Arthus-Bertrand (tl); Roger Wilmshurst; FLPA (bc). 203 Alamy Stock Photo: Loetscher Chlaus (tr). Corbis: Tony Hamblin; FLPA (cla). Dreamstime.com: Lunamarina (bl); Office2005 (cra). Getty Images / iStock: avideus (bc). 204 Alamy Stock Photo: RWP (cr). Dreamstime.com: Valentyn75 (bc). Getty Images / iStock: AlexeyMasliy (cla). Getty Images: Sabine Lubenow (tr). Science Photo Library: Nigel Cattlin (br). 204-205 Corbis: Chris Gibson. 205 Alamy Stock Photo: Mark Boulton (cr). Corbis: Chinch Gryzniewicz; Ecoscene (bc); Roger Tidman (tr). Chris Gibson: (bl). Neil Hardwick (cra). 206 Alamy Stock Photo: blickwinkel (cra). Corbis: Frank Blackburn, Ecoscene (cl); Raymond Gehman (br); Eric and David Hosking (cr); Robert Marien (cl). Getty Images: Ron Erwin (cra); Christopher Furlong (cr). 207 Alamy Stock Photo: Robography (clb). Corbis: Brandon D. Cole (cl); Roger Tidman (bl). Dorling Kindersley: Frank Greenaway courtesy of Natural History Museum, London (br); Colin Keates, courtesy of Natural History Museum, London (br). Getty Images: China Foto Press (cla). Chris Gibson: (t). Science Photo Library: Kjell B. Sandved (tr); Dr Keith Wheeler (c). Steve Trewhella (c). 208 Corbis: Paul A. Souders (c); Specialist Stock (c). 208-209 Getty Images: Johnny Johnson. 209 Alamy Stock Photo: Thomas Hanahoe (tr); Michael Patrick O'Neill (c); Rosanne Tackaberry (bc). Corbis: Karen Kasmauski / Science Faction (cb). Dreamstime.com: Klomsky (c). 212 Corbis: Galen Rowell (cr); Paul Nicklen / National Geographic (clb). 212-213 Corbis: Galen Rowell. 213 Alamy Stock Photo: Arcticphoto (c); Tom Ingram (tr). Dorling Kindersley: Frank Greenaway courtesy of Natural History Museum, London (br). Getty Images: Geoff du Feu (br). 214 Corbis: Steve Kaufman (br); Kennan Ward (cla). 214-215 naturepl.com: Konrad Wothe. 216 Alamy Stock Photo: Louise Murray (bl); WILDLIFE GmbH (tr). Dreamstime.com: Pär Edlund (tl). Getty Images / iStock: KenCanning (cl). Getty Images: Chris Jackson (tl). 216-217 Corbis: Bob Krist. Dreamstime.com: Adeliepenguin (tc). 217 Corbis: Momatiuk-Eastcott (cr); Paul Souders (cr). Dreamstime.com: Agami Photo Agency (tr)

All other images © Dorling Kindersley
For further information see: www.dkimages.com